Applications of Viscoelasticity

Applications of Viscoelasticity

Bituminous Materials Characterization and Modeling

POURIA HAJIKARIMI

Assistant Professor, Department of Civil & Environmental Engineering, Amirkabir University of Technology (Tehran Polytechnic), Tehran, Iran

FEREIDOON MOGHADAS NEJAD

Professor, Department of Civil & Environmental Engineering, Amirkabir University of Technology (Tehran Polytechnic), Tehran, Iran

ELSEVIER

Elsevier
Radarweg 29, PO Box 211, 1000 AE Amsterdam, Netherlands
The Boulevard, Langford Lane, Kidlington, Oxford OX5 1GB, United Kingdom
50 Hampshire Street, 5th Floor, Cambridge, MA 02139, United States

British Library Cataloguing-in-Publication Data
A catalogue record for this book is available from the British Library

Library of Congress Cataloging-in-Publication Data
A catalog record for this book is available from the Library of Congress

ISBN: 978-0-12-821210-3

For Information on all Elsevier publications
visit our website at https://www.elsevier.com/books-and-journals

Publisher: Matthew Deans
Acquisitions Editor: Dennis McGonagle
Editorial Project Manager: Ruby Smith
Production Project Manager: Kamesh Ramajogi
Cover Designer: Mark Rogers

Typeset by MPS Limited, Chennai, India

Dedication

Pouria Hajikarimi:
To *Sahar*

Fereidoon Moghadas Nejad:
This book is lovingly dedicated to my dearest, *Fariba*

Contents

Preface

There has always been a complicated impression for almost all graduate students in the field of civil engineering with the term "Viscoelasticity" as a fundamental theorem, especially for those who are enthusiastic about investigating bituminous materials characterization and modeling. Students are usually reluctant to learn theoretical subjects related to the viscoelasticity theorem and prefer to use it as a tool to interpret the results of experiments and run numerical simulations. There are several new and different aspects of the theoretical concepts and mathematical problems they face to understand the viscoelasticity theorem and then use to describe the rheological and mechanical behavior of materials like bitumen, asphalt mastic, asphalt mortar, or asphalt mixture, whose properties vary with time, temperature, and loading rate.

The idea of writing this book has come from the semester where the course of "Viscoelasticity Theorem" was presented for graduate students (both MSc and PhD students) of civil engineering in the field of pavement engineering for the first time. Reading fundamental textbooks and references may not always work for a graduate student searching to find the solution to a conventional and well-known viscoelastic problem. For example, interconversion of constitutive viscoelastic functions to determine the relaxation modulus of bituminous materials based on their creep compliance is a repetitive problem for defining material properties and developing numerical models. Several references described the general procedure for performing such an interconversion. Still, few of them presented a clear and straightforward solution that can be used by those who want to use the interconversion method quickly without spending much time studying and going into more depth. This book attempts to simply present routine viscoelasticity applications for bituminous materials characterization and modeling to be used by graduate students working in laboratories, experts in research and development departments of bitumen and asphalt industries, and all who want to have a brief and clear viewpoint on the viscoelasticity theorem and the widespread applications.

This book has been divided into eight chapters. The first three chapters are dedicated to an introduction and the basic concepts and fundamental formulations of the viscoelasticity theorem, which are necessary for those who intend to use this theory to understand the rheological and mechanical behavior of materials better. Two well-known main approaches are used to explain these essential concepts, including hereditary integration, by using the Boltzmann's superposition principle and differential equations considering both integer and fractional order equations. After that, Chapter 4, Correspondence Principle of Viscoelasticity, provides a brief and clear insight into the correspondence principle (CP) as an essential and useful method to derive analytical solutions of viscoelastic problems based on existing elastic solutions for both continuum and discontinuous media. The examples of using CP are presented in this chapter in detail and solved algorithmically to be conveniently used for similar problems.

The time–temperature superposition principle is presented in Chapter 5 as an important and brilliant topic of using viscoelasticity theory to characterize material properties. Chapter 5, Time–Temperature Superposition, is devoted to explaining the master curve concept, shift functions, and different master curve construction methods for simple thermorheological viscoelastic substances. Reading this chapter may help all those struggling with sketching a master curve for their experimental outputs at a specific reference temperature.

Since it is claimed in all relevant references and textbooks that by having one of the constitutive viscoelastic functions, relaxation modulus or creep compliance, it will be possible to determine another one, Chapter 6, Interconversion of Constitutive Viscoelastic Functions, is fully dedicated to introducing three different methods of interconversion previously established. To solve the practical examples of this chapter, which are all selected based on real experimental data, the MATLAB codes are provided, which one can efficiently use to interconvert one function to another one. While the presented MATLAB codes seem simple, we believe that they can efficiently help readers to understand the interconversion procedure better and assure them regarding their results.

Chapter 7, Application of Viscoelasticity for Experimental Tests, is devoted to introducing several types of experimental methods defined and analyzed based on the viscoelasticity theorem for characterizing the rheological and mechanical behavior of bituminous composites, including the three-point bending test, bending beam rheometer (BBR), semi-circular bending (SCB) test, and dynamic shear rheometer (DSR). Knowing the details of viscoelastic formulation for each test method can help both engineers and researchers check the results' meaningfulness and perform in-depth analysis.

Chapter 8, Incremental Form of Viscoelastic Relations, provides the incremental formulation for stress–strain analysis of viscoelastic materials in which detailed FEM MATLAB code is presented to solve a continuum and discontinuous media. Having a straightforward algorithm, it will be possible to use this incremental approach to determine a viscoelastic medium's stress or strain with a different type of loading and boundary conditions.

This book results from numerous brilliant and pioneer research works in the field of continuum mechanics, rheology, and viscoelasticity over many years all across the world. We have attempted to properly acknowledge the accomplishments of corresponding authors within the text, relevant figures, tables, and equations. We are much indebted to their outstanding research works and any unintended shortcoming in sufficiently acknowledging them is sincerely regretted.

We would like to extend our acknowledgment to Elsevier Publishing Company to trust us and facilitate the publication of this book; in particular Matthew Deans, Dennis McGonagle, Ruby Smith, and Kamesh Ramagoji for helping us throughout the work. We would also like to express our sincere gratitude to Dr. M.S. Kiasat, whose pamphlet of viscoelasticity theorem course facilitated writing Chapter 2, Constitutive Viscoelastic Functions, and Chapter 3, Mechanical Models of Viscoelasticity. Our special thanks go to our colleagues in the Department of Civil and Environmental Engineering at Amirkabir University of Technology for providing a calm and sincere atmosphere that allowed us to write this book during the past 2 years. Also, we want to thank all students who challenge us with their curiosity and helpful questions.

Finally, we would like to express our gratitude to our families for their love, understanding, and never-ending support. We have spent many hours writing this book, hours that could have been devoted to our families.

Pouria Hajikarimi and Fereidoon Moghadas Nejad
Tehran, Iran
Spring 2021

CHAPTER 1

Introduction to viscoelasticity

1.1 Introduction

Stresses, strains, and displacements are three types of quantities that are required to be determined in the mechanics of continuous media. The stresses, strains, and displacements are connected by three kinds of equations describing laws of nature: (1) equilibrium conditions, (2) kinematic relations, and (3) constitutive equations (Flügge, 2013). It is not possible to exactly explain the mechanical behavior of actual materials since they show a great variety of behavior. Therefore there is a common approach to finding the best idealized material model that can reflect the mechanical behavior of the desired material.

The elastic behavior (Hook's law) is one of the most straightforward material behaviors in which a one-to-one coordination exists between stress and strain (Fig. 1.1) (Sadd, 2009). Hook's law is widely used to describe the mechanical behavior of solid engineering materials, such as steel, concrete, coarse and fine aggregates, and fibers. An elastic material deforms instantaneously when stress is applied and remembers its original configuration. After removing the applied stress, it instantaneously returns to its original state. Several materials show the phenomenon of plastic flow which can be expressed by the following statements:

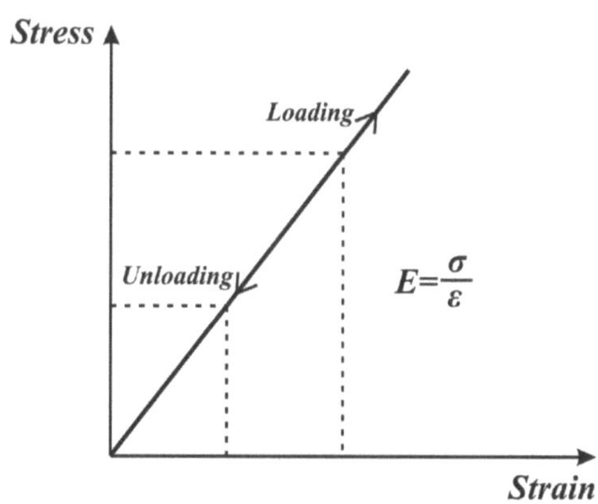

Figure 1.1 The elastic behavior (Hook's law).

Applications of Viscoelasticity
DOI: https://doi.org/10.1016/B978-0-12-821210-3.00008-5

1. The material behaves elastically until it reaches the yield limit.
2. Then, further strain is possible without increasing stress.
3. The additional strain is permanent and cannot be recovered after removing the applied stress.

The viscous behavior (Newton's law) is also used to explain the relation between shear stress and shear strain rate of a simple fluid. For Newtonian fluids that obey Newton's law, the ratio of shear stress to shear strain rate is constant and known as viscosity or coefficient of viscosity; a physical property to characterize the flow resistance of fluids. Fig. 1.2 schematically shows shear stress versus shear strain rate for a Newtonian fluid. As can be seen in this figure, it is evident that the coefficient of viscosity is not dependent on the shear rate but it may be dependent on temperature and pressure. Non-Newtonian fluids do not have constant viscosity as their viscosity is dependent on the shear rate, which may show shear-thinning or shear-thickening behavior, as shown in Fig. 1.2 (Currie, 2016). Unlike elastic materials, when the load is removed for a viscous material, the material has "forgotten" its original configuration, remaining in the deformed state.

Some materials exhibit both viscous and elastic properties when undergoing force load or deformation. A combinatorial name of "viscoelastic" was designated for such materials to show that their characteristics come from both the elastic and viscous idealized models. Viscoelasticity was observed in the late 20th century as a consequence of research works performed by physicists such as Maxwell (1867), Kelvin (1875), and Boltzmann in the 19th century on creep and recovery of rubbers, metals, and glasses. The origin of viscoelasticity can be found in polymer rheology

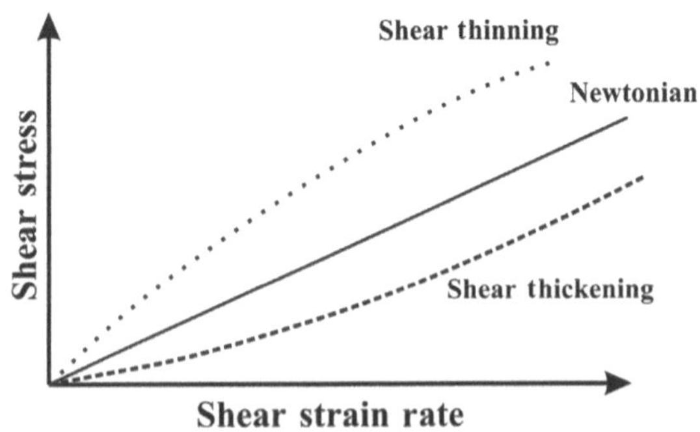

Figure 1.2 Shear stress versus shear strain rate for Newtonian and non-Newtonian's fluids.

(Brinson & Brinson, 2008) when synthetic polymers were designed, produced, and used in various applications in the late 20th century. Nevertheless, viscoelastic materials are not limited to polymers. Several engineering and industrial materials exhibit viscoelastic behavior, such as metals at very high temperatures (Fraizier et al., 2002), and bituminous materials like asphalt mastic, asphalt mortar, and asphalt mixture (Kim, 2009). The disks in the human spine, (Terry & Roberts, 1968), skin tissue (Silver et al., 2001), and wood (Ranta-Maunus, 1975) are classified as viscoelastic materials.

There is a prevalent point of view that viscoelasticity is far away from practical applications and design methods. Contrary to this opinion, neglecting viscoelastic characteristics of engineering materials like asphalt mixture leads to significant overestimation or underestimation in the procedure of designing the required parameters. For example, for designing flexible pavements, there are some old-fashioned solutions and methods, such as the two-layer or three-layer system of Burmister (1945), Burmister et al. (1944), that assumed all pavement layers to be elastic materials. Considering such an assumption neglects the dependency of an asphalt mixture's mechanical response to time and loading frequency and its ability for relaxing stresses resulted in overestimating or underestimating the thickness of an asphalt layer depending on climate conditions and traffic loading. Therefore it is important to recognize, evaluate, and analyze viscoelastic properties of considered materials to better understand their rheological and mechanical behavior.

The history of the viscoelasticity theorem returns to the topic of "fading memory" for materials subjected to time-dependent strain (Morrison, 2001). This behavior may be linear or nonlinear. When excited stress (or strain) is proportional to response strain (or stress), a material's behavior is linear, while for nonlinear behavior the response stress (or strain) is a function of applied strain (or stress).

Viscoelasticity is the consequence of the diffusion of molecules or atoms within an amorphous material. Therefore viscoelastic materials are characterized by considering time (frequency) of loading, temperature, and loading rate (Phan-Thien & Rossikhin, 2004). Furthermore, their response to deformation or load can be affected by previous deformation, load, or temperature history. Two common time-dependency forms of viscoelastic materials behavior are creep and stress relaxation, which are well-known in the technical literature for both experimental investigations and analytical derivations (Findley et al., 1976).

Creep is the propensity of a viscoelastic material to move gradually or deform permanently under the effect of a constant stress. Fig. 1.3 schematically depicts a strain diagram for a creep condition for a constant stress. The disks in the human spine creep under normal human body weight, and consequently they get shorter with time. This means that people are taller in the morning than in the evening! Therefore lying down allows these disks to recover and return to their former shape. For skin tissue as a viscoelastic material, one can observe creep behavior by pinching the skin at the back of the hand. After pinching, it takes time to recover to its original flat position and the required time for returning to the preliminary position depends on the time and intensity of pinching. A longer time of pinching results in a longer time of recovering, while rapid pinching leads to less time to recover (Lakes, 1998).

Stress relaxation is a gradual reduction in stress under a persistent deformation. Both these phenomena influence and, in many cases, limit the application of plastics for structural and load-bearing applications. Fig. 1.4 shows a schematic stress relaxation for a viscoelastic material. The strings of classical guitars used to be made of catgut (a natural polymer) and currently are made of nylon (a synthetic polymer). Both catgut and nylon are classified as viscoelastic materials. When a musician fixes the guitar strings at a constant length (keeping the strain constant), stress relaxation occurs. Then, the reduction in stress implies a reduction in frequency and a lowering of pitch and the guitar goes out of tune.

In general, given appropriate laboratory information on the behavior of viscoelastic materials under short-term and long-term stress or strain, it is possible for design engineers to overcome some of the difficulties of the

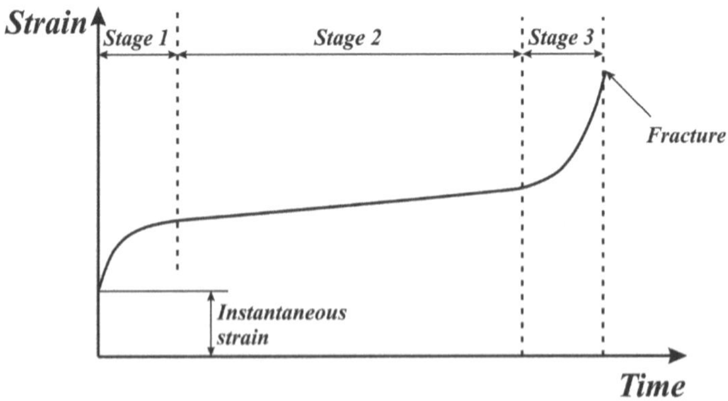

Figure 1.3 Schematic creep behavior of viscoelastic materials.

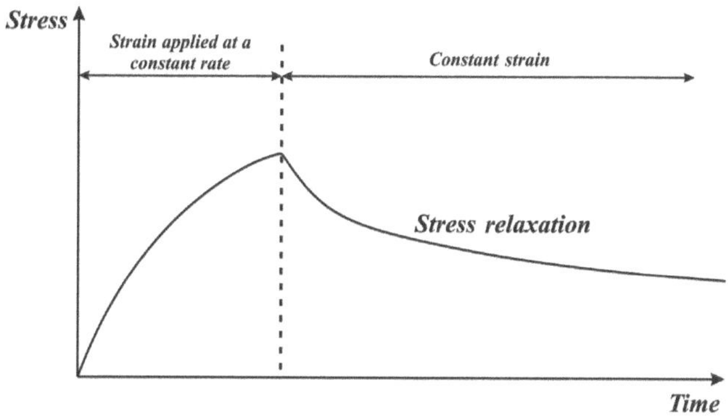

Figure 1.4 Schematic relaxation behavior of a viscoelastic material.

use of such materials under load-bearing conditions. The study of creep and stress relaxation phenomena is also of significant importance in evaluating fundamental viscoelastic functions of viscoelastic materials.

In the following, the context of each chapter presented in this book is briefly discussed to get more insight into the book's topics.

1.2 Constitutive viscoelastic functions

In order to classify materials based on an idealized theorem (elasticity, plasticity, viscoelasticity, hyperelasticity, etc.) for material modeling, it is required to define some special and representative parameters. For example, the Young's modulus and the Poisson's ratio are two main parameters for representing an elastic material. In the plasticity theorem, the characteristics' parameters are directly dependent on the yielding criteria (Lubliner, 2008). "Creep compliance" and "relaxation modulus" are two constitutive functions for representing a viscoelastic material and relating stress and strain together (Flügge, 2013). Generally, the creep compliance is defined as the ratio of time-dependent strain to time-dependent stress which can be kept constant to create a simple creep condition. Furthermore, the relaxation modulus is defined as the ratio of time-dependent stress to time-dependent strain which can be considered constant or linear to provide a simple stress relaxation condition. Correspondingly based on the abovementioned definitions, these functions can be determined by providing "creep" and "stress relaxation" conditions in the laboratory. One of the main challenges for laboratory experts and design engineers is to determine these functions accurately by implementing cost-effective apparatus. It is well-known to use Boltzmann's superposition principle (Brinson & Brinson, 2008) or mechanical models to derive constitutive equations for the relating stress and strain of a viscoelastic material having creep compliance or relaxation modulus, which are thoroughly presented in Chapter 2, Constitutive Viscoelastic Functions.

1.3 Mechanical models of viscoelasticity

There are two main approaches for modeling viscoelastic materials: (1) differential equations and (2) hereditary approach. The first approach contains integer-order and fractional-order differential equations, which easily make sense for students and those who first study the viscoelasticity theorem rather than the hereditary approach. In the integer-order differential equations approach, "springs" and "dashpots" are the two main components that represent elastic and viscous behavior, respectively (see Fig. 1.5). Different configurations of springs and dashpots connection resulted in developing different types of viscoelastic mechanical models. The oldest and simplest model was introduced by James Clerk Maxwell in 1867 by connecting a spring and a dashpot in series to find a constitutive equation for polymeric melts that can incorporate some elastic effects while still estimating viscous effects and flow. Then, other types of mechanical models were introduced by Kelvin (connecting a spring and a dashpot in parallel), Burgers, etc. Nowadays, the generalized Maxwell or

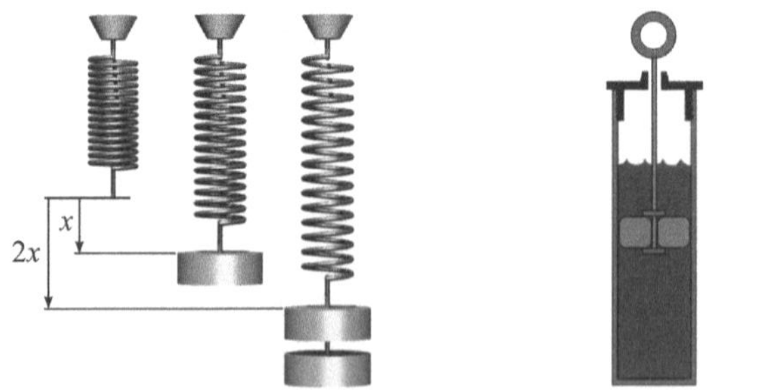

$$\sigma(t) = k\,\varepsilon(t) \qquad \sigma(t) = \eta\frac{d\varepsilon(t)}{dt}$$

Figure 1.5 Spring and dashpot as two main components of mechanical viscoelastic models.

Kevin models are widely used to characterize viscoelastic materials and introduce their properties to numerical frameworks.

Additionally, fractional-order differential equations are other types of mechanical viscoelastic models representing more complex behavior of viscoelastic materials based on fractional derivatives. The simple fractional element, generalized fractional viscoelastic model, Huet (1965), Huet-Sayegh (Pronk, 2006), 1S2P1D, and 2S2P1D (Olard & Di Benedetto, 2003) are some examples of fractional-order viscoelastic models, which are comprehensively introduced in Chapter 3, Mechanical Models of Viscoelasticity.

Both integer-order and fractional-order viscoelastic models determine some elastic and viscous parameters of real conceptual mechanical models. Therefore constitutive functions of creep compliance and relaxation modulus can be derived analytically for each mechanical model based on the relationship between spring and dashpot (or fractional dashpot) components. In order to using mechanical models, the main concern is the capability of the model to simulate the real viscoelastic behavior of the desired material. This is how some viscoelastic models are appropriate for simulating a special classification of considered viscoelastic material and other models are not.

1.4 The correspondence principle

As mentioned in Section 1.1, it is required to determine stresses, strains, and displacements as the three main quantities in the mechanics of continuous media. These parameters are exactly determined for some specific problems assuming the elastic behavior of all substances. Stress and strain distribution for an infinite half-space domain under a finite circular distributed load (Sadd, 2009), deflection of a three-point or four-point bending beam based on Euler–Bernoulli or Timoshenko theories (Todhunter, 2014), and displacement field near a crack tip are some examples of analytical elastic solutions having closed-form relations for determining stresses, strains, and displacements.

Having analytical solutions for the problems containing viscoelastic materials is also so attractive and it is of interest to engineers and researchers to calculate stresses, strains, and displacement for a test setup or real field loading conditions. It is not possible to directly apply elastic solutions on viscoelastic problems. The correspondence principle (CP) is manifested by using the Laplace transformation to find an analytical solution for a viscoelastic problem having its corresponding elastic solution with the same geometry, loading, and boundary conditions. The CP is introduced in Chapter 4, Correspondence Principle of Viscoelasticity, and different examples are used to provide more insight into using this principle for converting known elastic problems' solutions to the corresponding viscoelastic ones. However, it is not possible to find an analytical closed-form solution for all viscoelastic problems by applying the CP on their corresponding elastic solutions as inverse Laplace transformation cannot be determined straightforwardly for them.

1.5 Time–temperature–loading rate superposition

It is mentioned in Section 1.1 and it is well-known that the mechanical and rheological behaviors of viscoelastic materials are dependent on time, temperature, and loading rate. Fig. 1.6 simply shows the dependency of bitumen behavior (as a viscoelastic material) to temperature to flow. The flow behavior of bitumen could be the same for 1 h at 60°C or 10 h at 25°C. Therefore it can be roughly concluded that a longer time can be replaced by higher temperatures to achieve similar results.

From a general point of view, the time–temperature–loading rate superposition can be used to overcome some major experimental issues. For example, it is impossible to run some experiments at very high frequencies but this can be replaced with running the desired experiment at lower temperatures. Furthermore, some viscoelastic materials like bitumen cannot carry the load at higher temperatures (for example higher than 80°C for bitumen) in a conventional test setup but it can be tested at lower frequencies by using time–temperature superposition properties. Actually, the superposition principle is used to evaluate the mechanical time-dependent

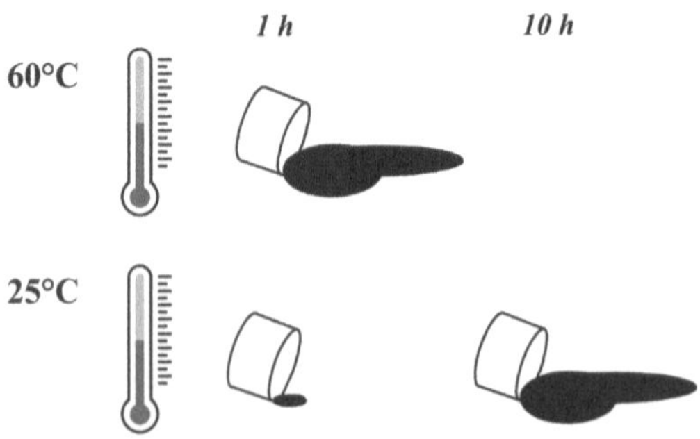

Figure 1.6 Flow behavior of bitumen depending on time and temperature.

properties of linear viscoelastic materials from known properties at a reference temperature (Ferry, 1980). For thermorheologically simple viscoelastic materials, the curves of the viscoelastic parameters like dynamic modulus, phase angle, relaxation modulus, and creep compliance as a function of time do not change shape as the temperature is altered but appear only to shift to the right or left. This suggests that a master curve at a specified temperature can be used as the reference to predict curves at different temperatures by applying a shift function. Several types of shift functions are introduced and implemented for constructing the master curve of viscoelastic materials in the technical literature, which are thoroughly presented and discussed in Chapter 5, Time–Temperature Superposition. Similar to the relation between time and temperature, a relationship can be found between time and temperature with loading rate, especially when viscoelastic materials exhibit nonlinear behavior.

1.6 Interconversion of constitutive viscoelastic functions

Theoretically, knowing one of the constitutive viscoelastic functions (creep compliance or relaxation modulus) makes it possible to determine another one without performing additional experimental tests (Schwarzl, 1970). Sometimes, one may need a viscoelastic function to use it in a numerical framework or capture a specific parameter but measuring another constitutive function is easier in the laboratory. For example, determination of creep compliance for bituminous materials is more accurate and easier than the relaxation modulus in the laboratory but the relaxation modulus is required as an input material property in well-known numerical simulator software like ANSYS or ABAQUS. Therefore it is prevalent to measure the creep compliance in the laboratory and then interconvert it to the relaxation modulus for use in the desired numerical framework.

To practically interconvert a constitutive viscoelastic function, several methods have been introduced and used in the technical literature. Some methods are based on a step-by-step solution, like the method presented by Hopkins and Hamming. Another method is based on solving a system of algebraic equations considering the Prony series for representing the creep compliance or the relaxation modulus of viscoelastic materials. Additionally, there are several approximation methods existing in the technical literature, such as methods introduced by Schapery (1962), Booij and Thoone (1982), and Christensen (2012). Some researchers

(Hajikarimi et al., 2018; Underwood et al., 2010) also used a particular function like a general power-law to derive an analytical relationship for interconverting constitutive viscoelastic functions. Interconversion methods are introduced and formulated in Chapter 6, Interconversion of Constitutive Viscoelastic Functions, and some practical examples are supported with MATLAB® codes.

1.7 Application of viscoelasticity for experimental tests

Despite developing a robust framework for defining constitutive functions and developing mechanical models to describe the behavior of viscoelastic materials, it is required to establish reliable and replicable experiments to evaluate these parameters accurately. Since rheological and mechanical behavior of viscoelastic materials is dependent on time, temperature, and loading rate, using an appropriate experimental test considering suitable conditions is a major concern for achieving the desired characteristics of viscoelastic materials. For example, according to ASTM D7175-15, the dynamic shear rheometer (DSR) device is commonly used to measure complex shear modulus and phase angle of bituminous materials at moderate and high test temperatures with two different sizes of specimens, as shown in Fig. 1.7. This figure shows two bitumen specimens with diameters of 8 and 25 mm that are used to perform shear rheometry at moderate (approximately 10°C–40°C) and high test temperatures (approximately 40°C–90°C), respectively. Fig. 1.8 also shows two geometries at the height of these specimens called plate–plate and cone–plate which have different stress distributions.

It is evident that measuring the characteristics of viscoelastic materials is subject to several challenges, such as controlling test temperature, performing the experiment at desired frequency, and applying different

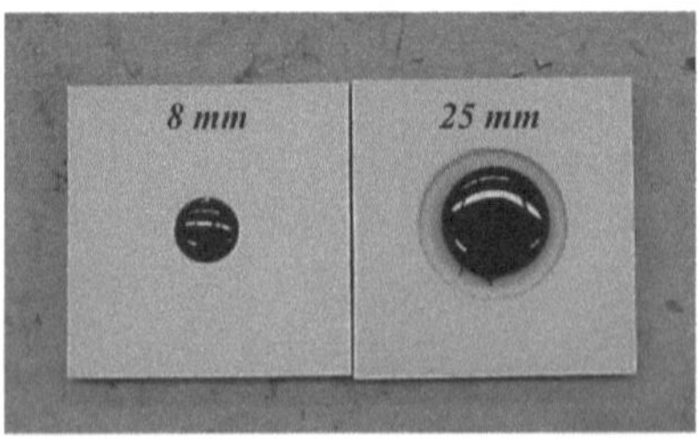

Figure 1.7 Two different diameters of specimens for performing experiments with dynamic shear rheometer (DSR).

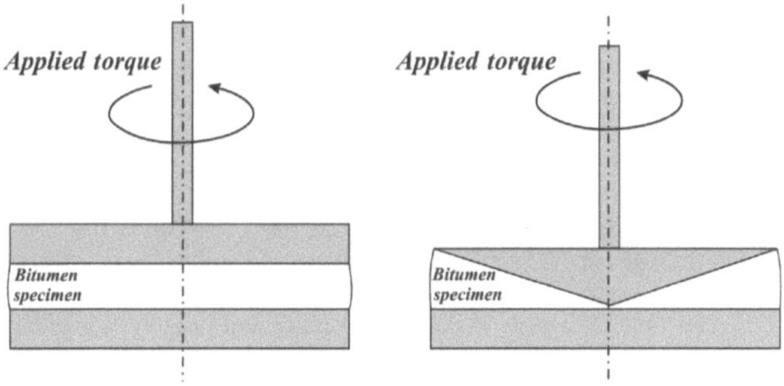

Figure 1.8 Two different geometries of specimens, parallel–plate and cone–plate, for performing experiments with dynamic shear rheometer (DSR).

loading types including creep, stress relaxation, frequency sweep, time sweep, temperature sweep, strain sweep, repeated creep and recovery (Bahia et al., 2001), multiple stress creep and recovery (ASTM D7405-20), and linear amplitude sweep (AASHTO TP 101-14). Chapter 7, Application of Viscoelasticity for Experimental Tests, presented different test methods for characterizing bituminous composites as viscoelastic materials and some Excel and MATLAB codes are provided to simply analyze typical test results to determine the desired parameters.

1.8 Incremental form of viscoelastic relations

How it is possible to apply viscoelastic constitutive relations for numerically solving a problem to determine stresses, strains, and displacements of a viscoelastic domain? It is possible to express this question in another way: how it is possible to involve viscoelastic constitutive relations within the finite element method (FEM) framework or other well-known numerical methods like finite difference method or extended finite element method (XFEM)? Since viscoelastic relations are dependent on time, an incremental formulation is generally used to determine time-dependent stresses or strains step-by-step considering appropriate time steps. Chapter 8, Incremental Form of Viscoelastic Relations, introduces such a formulation and uses it to solve different types of viscoelastic problems by providing required functions written in MATLAB.

References

Bahia, H.U., Hanson, D.I., Zeng, M., Zhai, H., Khatri, M.A., & Anderson, R.M. (2001). *Characterization of modified asphalt binders in superpave mix design* (Issue Project 9-10 FY'96).

Booij, H. C., & Thoone, G. P. J. M. (1982). Generalization of Kramers-Kronig transforms and some approximations of relations between viscoelastic quantities. *Rheologica Acta, 21*(1), 15–24. Available from https://doi.org/10.1007/BF01520701.

Brinson, H. F., & Brinson, L. C. (2008). *Polymer engineering science and viscoelasticity: An introduction*. Springer.

Burmister, D. M. (1945). The general theory of stresses and displacements in layered systems. I. *Journal of Applied Physics, 16*(2), 89–94.

Burmister, D. M., Palmer, L. A., Barber, E. S., & Middlebrooks, T. A. (1944). The theory of stress and displacements in layered systems and applications to the design of airport runways. *Highway Research Board Proceedings, 23*.

Christensen, R. (2012). *Theory of viscoelasticity: An introduction*. Elsevier.

Currie, I. G. (2016). *Fundamental mechanics of fluids*. CRC Press.

Ferry, J. D. (1980). *Viscoelastic properties of polymers*. John Wiley & Sons.

Findley, W.N., Lai, J.S., & Onaran, K. (1976). *Creep and relaxation of nonlinear materials*. North-Holland, Amsterdam.

Flügge, W. (2013). *Viscoelasticity*. Springer Science & Business Media.

Fraizier, E., Nadal, M.-H., & Oltra, R. (2002). Evaluation of viscoelastic constants of metallic materials by laser-ultrasonics at elevating temperature. *Ultrasonics, 40*(1), 543–547. Available from https://doi.org/10.1016/S0041-624X(02)00165-8.

Hajikarimi, P., Nejad, F. M., & Aghdam, M. M. (2018). Implementing general power law to interconvert linear viscoelastic functions of modified asphalt binders. *Journal of Transportation Engineering Part B: Pavements, 144*(2). Available from https://doi.org/10.1061/JPEODX.0000038.

Huet, C. (1965). *Etude par une méthode d'impédance du comportement viscoélastique des matériaux hydrocarbonés.*

Kelvin, W. (1875). On the theory of viscoelastic fluids. *Mathematical and Physical Papers, 3*, 27–84.

Kim, R. Y. (2009). *Modeling of asphalt concrete*. Mc-Graw Hill.

Lakes, R. S. (1998). *Viscoelastic solids* (Vol. 9). CRC Press.

Lubliner, J. (2008). *Plasticity theory*. Courier Corporation.

Maxwell, J. C. (1867). On the dynamical theory of gases. *Philosophical Trans- actions Royal Society, A157*, 49–88.

Morrison, F. A. (2001). *Understanding rheology*. United States: Oxford University Press.

Olard, F., & Di Benedetto, H. (2003). General "2S2P1D" model and relation between the linear viscoelastic behaviours of bituminous binders and mixes. *Road Materials and Pavement Design, 4*(2), 185–224.

Phan-Thien, N., & Rossikhin, Y. A. (2004). Understanding viscoelasticity: Basics of rheology. *Applied Mechanics Reviews, 57*(1), B4, B4.

Pronk, A.C. (2006). The Huet-Sayegh model: A simple and excellent rheological model for master curves of asphaltic mixes. In *Asphalt concrete: Simulation, modeling, and experimental characterization* (pp. 73–82).

Ranta-Maunus, A. (1975). The viscoelasticity of wood at varying moisture content. *Wood Science and Technology, 9*(3), 189–205.

Sadd, M. H. (2009). *Elasticity: Theory, applications, and numerics*. Academic Press.

Schapery, R. (1962). Approximate methods of transform inversion for viscoelastic stress analysis. *Proceedings of the fourth U.S. National Congress of Applied Mechanics, 2*.

Schwarzl, F. R. (1970). On the interconversion between viscoelastic material functions. *Pure and Applied Chemistry*, *23*(2–3), 219–234.
Silver, F. H., Freeman, J. W., & DeVore, D. (2001). Viscoelastic properties of human skin and processed dermis. *Skin Research and Technology*, 7(1), 18–23.
Terry, C. T., & Roberts, V. L. (1968). A viscoelastic model of the human spine subjected to + gz accelerations. *Journal of Biomechanics*, *1*(2), 161–168.
Todhunter, I. (2014). *A history of the theory of elasticity and of the strength of materials*. Cambridge University Press.
Underwood, B., Far, M., & Kim, Y. (2010). Using limited purchase specification tests to perform full linear viscoelastic characterization of asphalt binder. *Journal of Testing and Evaluation*, *38*(5), 558–566. Available from https://doi.org/10.1520/JTE102591.

CHAPTER 2

Constitutive viscoelastic functions

There are two well-known constitutive functions for describing the mechanical behavior of viscoelastic materials, which is dependent on time, temperature, and loading rate: creep compliance and relaxation modulus. Creep compliance is the strain developed during a creep test when a unit stress is applied. Relaxation modulus is the stress developed during a relaxation test when a unit strain is applied. These two functions can properly reflect viscoelastic material properties and interconvert to each other. Therefore it is required to at least determine one of these two functions for characterizing the desired viscoelastic material by designing and performing standard experimental tests such as creep test, relaxation test, or constant strain-rate test (Ottosen & Ristinmaa, 2005). Having creep compliance or relaxation modulus, it will be possible to establish the stress—strain relation of a viscoelastic material which can work for different loading conditions. In this chapter, the constitutive viscoelastic functions are defined and the hereditary approach is used to develop the stress—strain relation for a linear viscoelastic material.

2.1 Creep compliance

The creep test is a simple method to determine the creep compliance of a viscoelastic material. Applying a constant stress over a specific time and capturing the resulting strain is the main concept of this test to determine creep compliance based on Formula (2.1):

$$D(t) = \frac{\varepsilon(t)}{\sigma_0} \tag{2.1}$$

in which $\varepsilon(t)$ is the captured strain and σ_0 is the stress applied instantaneously and then kept constant. Fig. 2.1 depicts the schematic stress and strain history of a creep test in which t_1 is the required time for achieving the required stress in the step-stress experiment. Fig. 2.2 also shows a real creep test on bitumen in which an instantaneous load of 0.98 N is applied at the mid-span of a three-point bending beam and then held constant for

Applications of Viscoelasticity
DOI: https://doi.org/10.1016/B978-0-12-821210-3.00002-4

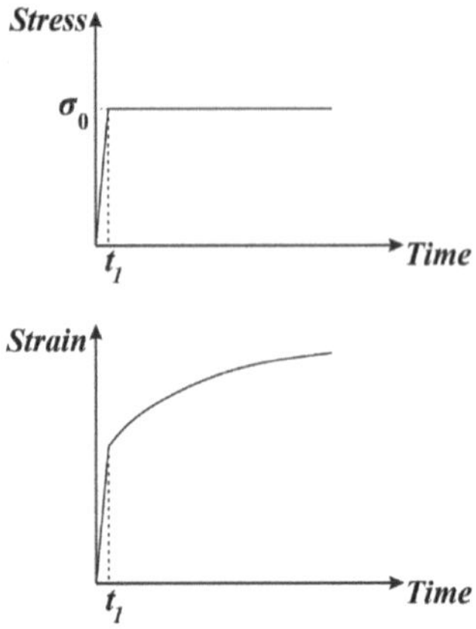

Figure 2.1 Schematic stress and strain history of a creep test.

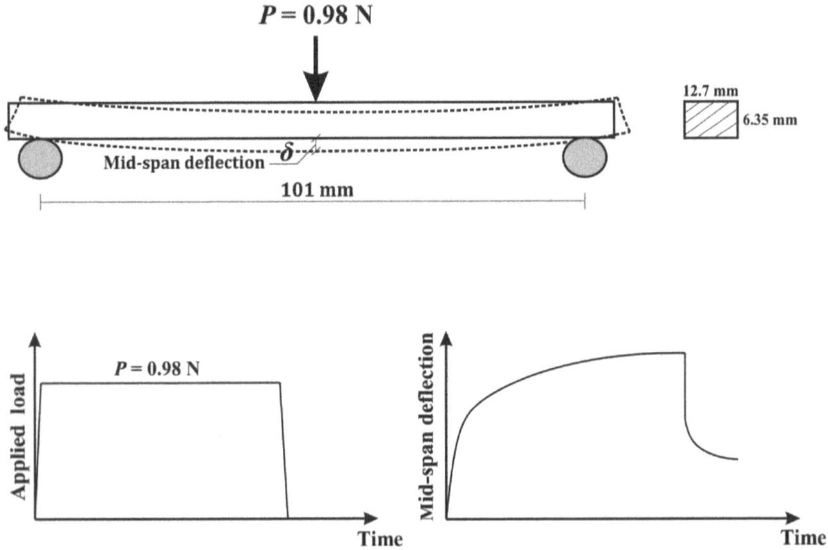

Figure 2.2 Bending beam rheometer (BBR) as a three-point creep bending test for determining creep compliance of bitumen.

240 s at low temperatures. The strain history is captured by measuring the beam deflection under the point of loading in 0.5 s time steps. It is a well-known test of the bending beam rheometer (BBR) (ASTM D6648-08, 2016, Standard Test Method for Determining the Flexural Creep Stiffness of Asphalt Binder Using the Bending Beam Rheometer (BBR), 2016) for determining the flexural creep stiffness of bitumen which is

defined as the inverse of creep compliance. The flexural creep stiffness, $S(t)$, and the creep compliance of bitumen, $D(t)$, are determined as follows:

$$S(t) = \frac{PL^3}{4bh^3\delta(t)} \tag{2.2}$$

$$D(t) = \frac{1}{S(t)} \tag{2.3}$$

where P is the applied constant load of 0.98 N at the mid-span of beam, L the span length of 101 mm, b the beam width of 12.7 mm, h the beam height of 6.35 mm, and $\delta(t)$ is mid-span deflection of beam (mm).

Fig. 2.3 shows the results of a BBR test performed on an original bitumen specimen at different test temperatures. As can be seen in this figure, it is evident that increasing the test temperature and time of loading increases the mid-span deflection and consequently creep compliance of the bitumen. The creep compliance as a constitutive viscoelastic function is dependent on time and temperature.

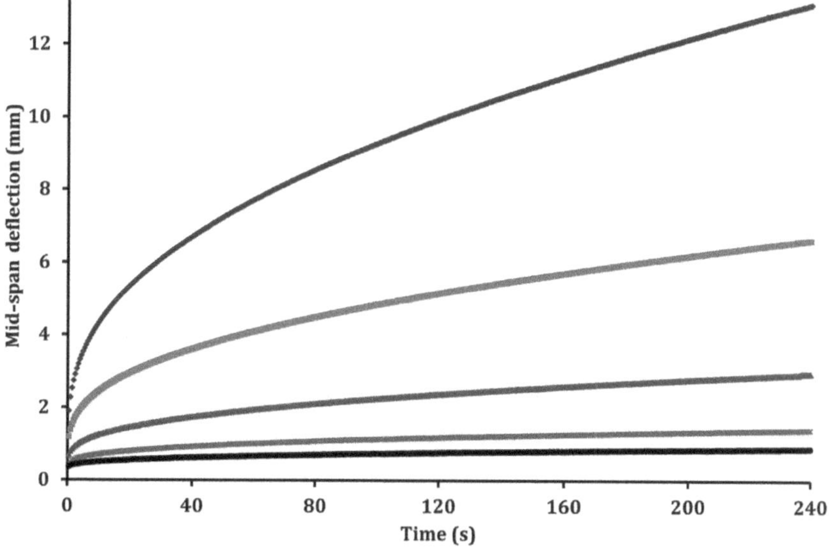

Figure 2.3 Results of a BBR test on an original bitumen at different test temperatures of −6°C, −12°C, −18°C, −24°C, and −30°C.

2.2 Relaxation modulus

The relaxation test is a well-known method to measure the relaxation modulus of a viscoelastic material. Applying a constant strain during a specific time and capturing the resulting stress is the basis of this test to determine the relaxation modulus based on Eq. (2.4):

$$E(t) = \frac{\sigma(t)}{\varepsilon_0} \tag{2.4}$$

in which $\sigma(t)$ is the measured stress and ε_0 is the applied strain which is kept constant during the test procedure. Fig. 2.4 depicts the schematic strain and stress history of a relaxation test in which t_1 is the required time for achieving the required strain in the step-strain experiment. It is necessary to mention that the application of the strain should be performed suddenly without any dynamic or intertia effects. It is a common method that the specimen is suddenly stretched to a new status and rigidly fixed such that the strain remains constant for the duration of the test.

As can be seen in Fig. 2.4, there are two types of relaxation behavior for viscoelastic materials: thermoplastic and thermoset. For an ideal thermoplastic polymer, the stress will tend to zero but for a thermoset polymer, it will decrease to a constant value and will not reach to zero, which is called residual stress (Brinson & Brinson, 2008). Fig. 2.5 also depicts the relaxation modulus of unaged asphalt binder at different test temperatures. It is evident that by increasing time and temperature of the test, the

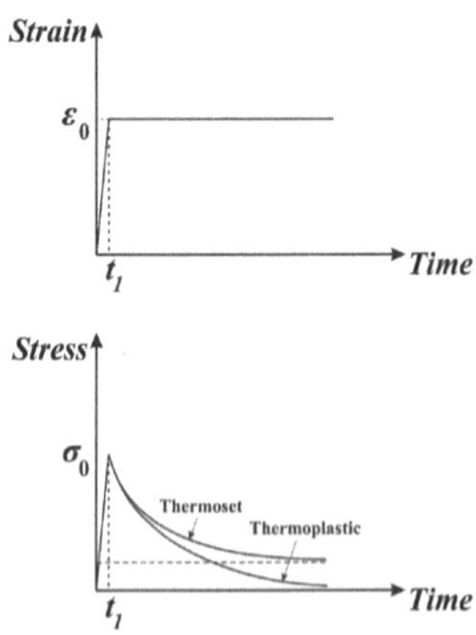

Figure 2.4 Schematic strain and stress history of a relaxation test.

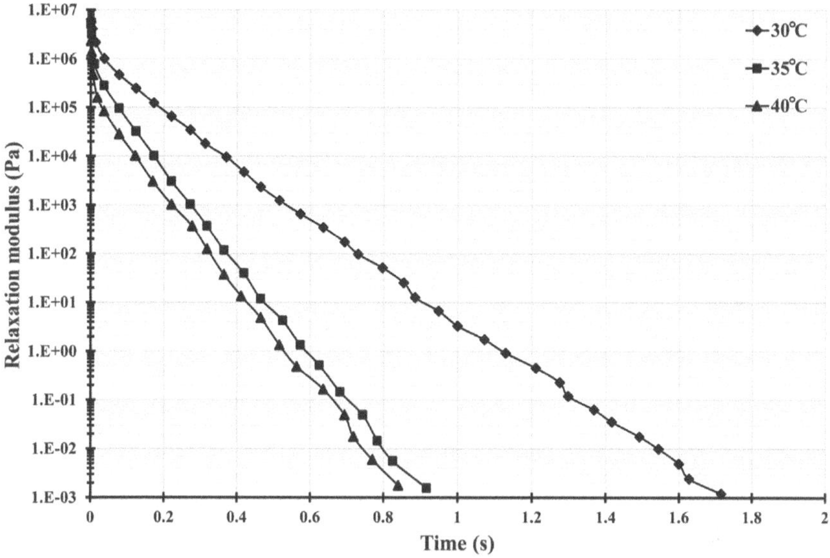

Figure 2.5 Relaxation modulus of an unaged asphalt binder at different test temperatures of 30°C, 35°C, 40°C, and 45°C. *From Padmarekha, A., & Murali Krishnan, J. (2013). Viscoelastic transition of unaged and aged asphalt. Journal of Materials in Civil Engineering, 25(12), 1852–1864. https://doi.org/10.1061/(ASCE)MT.1943-5533.0000734.*

relaxation modulus decreases, which again reminds one that viscoelastic properties depend on time and temperature.

2.3 Boltzmann's superposition principle

As mentioned earlier, the hereditary approach is introduced in this chapter to determine constitutive viscoelastic functions for a linear viscoelastic material. The vital issue for applying the hereditary approach is that of superposition which is possible due to assuming linear behavior for viscoelastic materials as a basic assumption in this chapter.

An arbitrary uniaxial stress history is shown in Fig. 2.6A as an input excitation function. By dividing the whole time of stress application into n small time steps, it will be an acceptable assumption that the applied stress is considered to be a constant amount of $\Delta\sigma$ in each time step. Having creep compliance of a viscoelastic material, $D(t)$, it is possible to determine the corresponding strain $\Delta\varepsilon(t)$ at time ξ by using Eq. (2.5) as follows:

$$\Delta\varepsilon(t) = D(t-\xi)\Delta\sigma \tag{2.5}$$

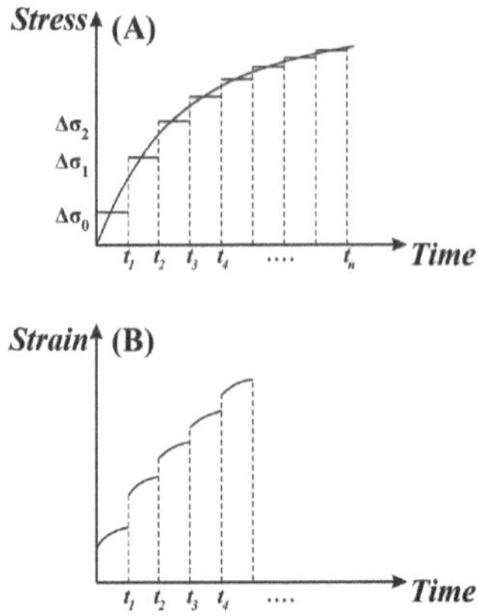

Figure 2.6 Boltzmann's superposition for (A) an arbitrary uniaxial stress history divided into *n* time steps and (B) corresponding strain for each time step.

for each time step as it is depicted in Fig. 2.6B. It is possible to sum the corresponding strains over the entire loading history (*n* time steps) as:

$$\varepsilon(t) = \sum_{i=0}^{n} \left[\Delta\sigma_i D(t-\xi_i)\right] \times \frac{\Delta\xi_i}{\Delta\xi_i} = \sum_{i=0}^{n} \frac{\Delta\sigma_i}{\Delta\xi_i} D(t-\xi_i) \times \Delta\xi_i \quad (2.6)$$

By considering an infinitesimal stress change $d\sigma$ applied at time ξ, Eq. (2.6) can be written in a differential form as:

$$\varepsilon(t) = \int_{-\infty}^{t} D(t-\xi)\frac{d\sigma(\xi)}{d\xi}d\xi \quad (2.7)$$

if the stress history $\sigma = \sigma(\xi)$ is smooth. However, if the stress history exhibits jumps, then integration of Eq. (2.8) should be used.

$$\varepsilon(t) = \int_{-\infty}^{t} D(t-\xi)d\sigma(\xi) \quad (2.8)$$

The lower integration limit in hereditary integral constitutive equations is taken as $-\infty$ by tradition. By supposing an instantaneous loading occurs at time zero according to:

$$\sigma(t) = \begin{cases} 0 & \text{when } t < 0 \\ \sigma_0 + \sigma_1(t); \quad \sigma_1(0) = 0 & \text{when } t \geq 0 \end{cases} \quad (2.9)$$

it is possible to determine the strain by discretizing integration limits as:

$$\begin{aligned}\varepsilon(t) &= \int_{-\infty}^{0^-} D(t-\xi)d\sigma(\xi) + \int_{0^-}^{0^+} D(t-\xi)d\sigma(\xi) + \int_{0^+}^{t} D(t-\xi)d\sigma(\xi) \\ &= 0 + D(t)\sigma_0 + \int_{0^+}^{t} D(t-\xi)d\sigma(\xi)\end{aligned} \tag{2.10}$$

i.e.,

$$\varepsilon(t) = D(t)\sigma_0 + \int_{0^+}^{t} D(t-\xi)d\sigma(\xi) \tag{2.11}$$

Fig. 2.7 shows an arbitrary uniaxial strain excitation function and its corresponding stress response. Similar to the procedure that was applied for determining strain by having stress excitation function, by dividing the entire time of strain application into n small time steps, it will be an appropriate assumption that the applied strain is considered a constant amount of $\Delta\varepsilon$ in each time step. Having a relaxation modulus of a viscoelastic material, $E(t)$, the corresponding stress, $\Delta\sigma(t)$, can be determined at time ξ by using Eq. (2.12) as follows:

$$\Delta\sigma(t) = E(t-\xi)\Delta\varepsilon \tag{2.12}$$

The corresponding stresses over the whole strain history (n time steps) can be accumulated as:

$$\sigma(t) = \sum_{i=0}^{n} \left[\Delta\varepsilon_i E(t-\xi_i)\right] \times \frac{\Delta\xi_i}{\Delta\xi_i} = \sum_{i=0}^{n} \frac{\Delta\varepsilon_i}{\Delta\xi_i} E(t-\xi_i) \times \Delta\xi_i \tag{2.13}$$

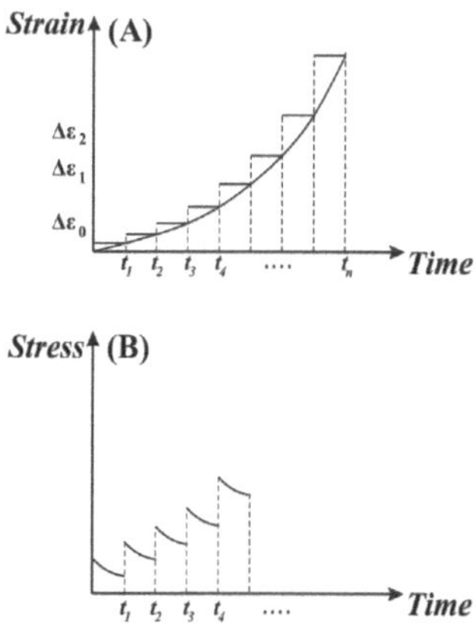

Figure 2.7 Boltzmann's superposition for (A) an arbitrary uniaxial strain history divided into *n* time steps and (B) corresponding stress for each time step.

and by assuming an infinitesimal strain change $d\varepsilon$ applied at time ξ, Eq. (2.13) can be written in a differential form as:

$$\sigma(t) = \int_{-\infty}^{t} E(t-\xi)\frac{d\varepsilon(\xi)}{d\xi}d\xi \tag{2.14}$$

for a smooth strain history $\varepsilon = \varepsilon(\xi)$. However, if the strain history is not smooth and exhibits some jumps, then the following integration should be used.

$$\sigma(t) = \int_{-\infty}^{t} E(t-\xi)d\varepsilon(\xi) \tag{2.15}$$

Similar to the derivation of Eq. (2.11), it is possible to rewrite Eq. (2.15) as follows:

$$\sigma(t) = E(t)\varepsilon_0 + \int_{0^+}^{t} E(t-\xi)d\varepsilon(\xi) \tag{2.16}$$

2.4 Response of viscoelastic material to harmonic loading

One of the most common excitations which can be applied for capturing linear or nonlinear characteristics of a viscoelastic material is harmonic loading, as shown in Fig. 2.8, using a dynamic mechanical analyzing (DMA) setup, and can be described as follows:

$$\gamma(t) = \gamma_0 \sin(\omega t) \tag{2.17}$$

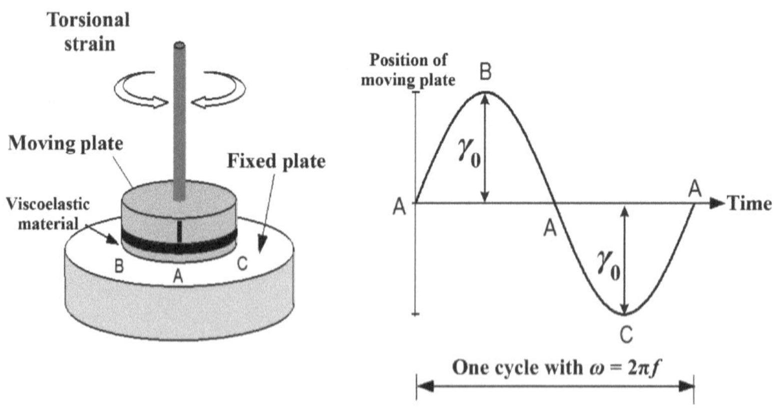

Figure 2.8 Dynamic mechanical analyzer for applying harmonic load on viscoelastic material.

in which γ_0 is the excitation amplitude of shear strain and ω is the excitation frequency which shows loading rate. It is expected that the response shear stress has the following form:

$$\tau(t) = \tau_0 \sin(\omega t - \varphi) \tag{2.18}$$

where φ is the time lag between the application of strain and stress response (see Fig. 2.9). This parameter is called "phase angle" in the viscoelastic literature. It is evident that having $\varphi = 0°$, we have ideal elastic behavior, and having $\varphi = 90°$, we have ideal viscous behavior. Since the excitation function is smooth and continuous, it is possible to use the hereditary integral constitutive equation of Eq. (2.14) for determining response stress by substituting the derivative of strain function in the abovementioned equation as:

$$\tau(t) = \int_{-\infty}^{t} G(t - \xi)\gamma_0 \omega \cos(\omega \xi) d\xi \tag{2.19}$$

where $G(t)$ is shear relaxation modulus. Changing the integration variable as $\eta = t - \xi$ and knowing that $d\eta = -d\xi$, Eq. (2.19) can be simply rewritten as:

$$\sigma(t) = \int_{\infty}^{0} G(\eta)\gamma_0 \omega \cos[\omega(t - \eta)](-d\eta) = \int_{0}^{\infty} G(\eta)\gamma_0 \omega \cos[\omega(t - \eta)] d\eta \tag{2.20}$$

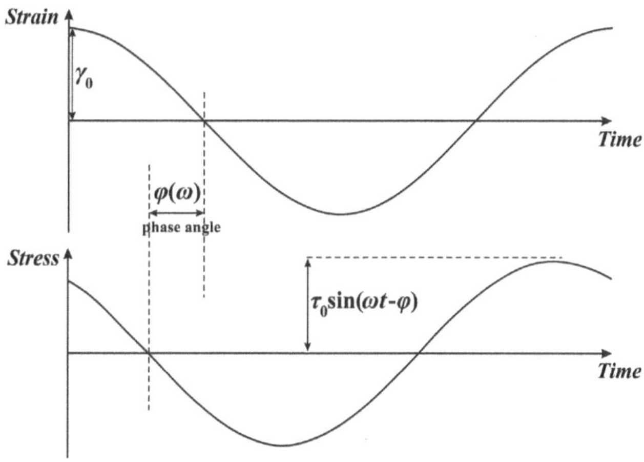

Figure 2.9 Strain input and stress output for a steady-state harmonic loading. *From Brinson, H. F., & Brinson, L. C. (2008).* Polymer engineering science and viscoelasticity, an introduction. *Springer. 186.*

Remembering the expansion formula for cos(a − b) = cos(a)cos(b) + sin(a)sin(b), it simply can be shown that:

$$\cos[\omega(t-\eta)] = \cos(\omega t)\cos(\omega\eta) + \sin(\omega t)\sin(\omega\eta) \tag{2.21}$$

Then Eq. (2.20) can be expanded as follows:

$$\tau(t) = \gamma_0 \int_0^{\infty} [\omega G(\eta)\cos(\omega\eta)d\eta]\cos(\omega t) + \gamma_0 \int_0^{\infty} [\omega G(\eta)\sin(\omega\eta)d\eta]\sin(\omega t) \tag{2.22}$$

It is possible to reorganize Eq. (2.22) as:

$$\tau(t) = \gamma_0[G'(\omega)\sin(\omega t) + G''(\omega)\cos(\omega t)] \tag{2.23}$$

where $G'(\omega)$ is storage modulus which measures the stored energy and represents the elastic portion (in phase with the applied strain) and $G''(\omega)$ is loss modulus which measures the energy dissipated as heat and represents the viscous portion (perpendicular to the applied strain). These modulus can be represented by comparing Eq. (2.23) with Eq. (2.22) as:

$$G'(\omega) = \int_0^{\infty} \omega G(\eta)\sin(\omega\eta)d\eta \tag{2.24}$$

$$G''(\omega) = \int_0^{\infty} \omega G(\eta)\cos(\omega\eta)d\eta \tag{2.25}$$

where ∞ in the upper integration limit denotes the time required to reach the steady-state condition for the response of the viscoelastic material.

There is another form to represent storage and loss moduli of a viscoelastic material by using complex mathematics and considering the real portion of the excited strain which can be defined as:

$$\gamma(t) = \Re\left(\gamma_0 e^{i\omega t}\right) = \gamma_0\cos(\omega t) \tag{2.26}$$

and having the relation of $G^*(i\omega) = [G'(\omega) + iG''(\omega)]$ for dynamic modulus as a combination of storage and loss moduli in the complex domain, the response stress can be represented as follows:

$$\begin{aligned}\tau(t) &= \Re\left[\gamma_0 G^*(i\omega)e^{i\omega t}\right] = \Re\left\{\gamma_0[G'(\omega) + iG''(\omega)][\cos(\omega t) + i\sin(\omega t)]\right\} \\ &= \gamma_0[G'(\omega)\cos(\omega t) - G''(\omega)\sin(\omega t)]\end{aligned} \tag{2.27}$$

The dynamic modulus can be represented as a vector in the complex plane, as depicted in Fig. 2.10. Therefore it can be simply found that:

$$G^*(i\omega) = G'(\omega) + iG''(\omega) = \left|G^*(i\omega)\right| e^{i\delta(\omega)} \tag{2.28}$$

where

$$\left|G^*(i\omega)\right| = \sqrt{(G'(\omega))^2 + (G''(\omega))^2} \tag{2.29}$$

Regarding to Fig. 2.8, it is possible to determine $\tan(\delta)$, which sometimes is called dissipated energy ratio, as:

$$\tan[\delta(\omega)] = \frac{G''(\omega)}{G'(\omega)} \tag{2.30}$$

For harmonic stress excitation, as $\tau(t) = \tau_0 \sin(\omega t)$, by using a similar procedure, it is possible to determine storage compliance, $J'(\omega)$, and loss compliance, $J''(\omega)$, as follows:

$$J'(\omega) = \int_0^{\infty} \omega J(\zeta) \sin(\omega\zeta) d\zeta \tag{2.31}$$

$$J''(\omega) = \int_0^{\infty} \omega J(\zeta) \cos(\omega\zeta) d\zeta \tag{2.32}$$

where ∞ in the upper integration limit is the time needed to achieve the steady-state condition for the response of the viscoelastic material. Similar

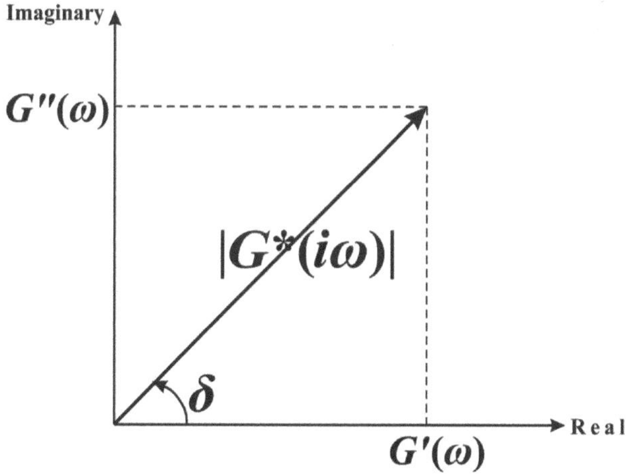

Figure 2.10 Storage and loss modulus in the complex plane.

to defining the dynamic modulus, the dynamic compliance can be defined in the complex domain by a combination of storage compliance and loss compliance as follows:

$$J^*(i\omega) = \left[J'(\omega) + iJ''(\omega)\right] e^{i\delta(\omega)} \tag{2.33}$$

References

ASTM D6648-08(2016). (6648). *Standard test method for determining the flexural creep stiffness of asphalt binder using the bending beam rheometer (BBR).* West Conshohocken, PA: ASTM International. Available from https://doi.org/10.1520/D6648-08R16.

Brinson, H. F., & Brinson, L. C. (2008). *Polymer engineering science and viscoelasticity, an introduction.* Springer.

Ottosen, N. S., & Ristinmaa, M. (2005). *The mechanics of constitutive modeling.* Elsevier.

CHAPTER 3

Mechanical models of viscoelasticity

3.1 Introduction

In addition to the hereditary approach for describing and evaluating the mechanical behavior of viscoelastic materials, there is another approach to use simple mechanical models like spring and dashpot and their combinations to explain the stress–strain relation of viscoelastic materials. In this chapter, different viscoelastic mechanical models are represented, including two main categories of integer-order and fractional-order models. The first category contains conventional and old-fashioned models like the Maxwell model, the Kelvin–Voigt model, the Burgers model, and generalized Maxwell and Kelvin–Voigt models, while the second category includes the fractional ones such as Huet–Sayegh.

3.2 Integer-order differential equations

There are two main components for developing integer-order differential equations: linear spring and dashpot. All integer-order differential equations used to describe the mechanical behavior of viscoelastic materials are constructed by various combinations of these two components. It is well-known that linear spring and dashpot behave as a pure elastic (Hooke's law) and viscous (Newton's law) material, respectively. Fig. 3.1 shows linear spring and dashpot and their corresponding stress–strain relations represented in Eqs. (3.1) and (3.2).

$$\sigma_e = E\varepsilon_e \tag{3.1}$$

$$\sigma_v = \eta\dot{\varepsilon}_v \tag{3.2}$$

In which σ and ε are applied stress and captured strain, E is stiffness of the linear spring, η is viscosity coefficient of the dashpot, and subscripts e and v refer to elastic and viscous behavior, respectively.

Applications of Viscoelasticity
DOI: https://doi.org/10.1016/B978-0-12-821210-3.00003-6

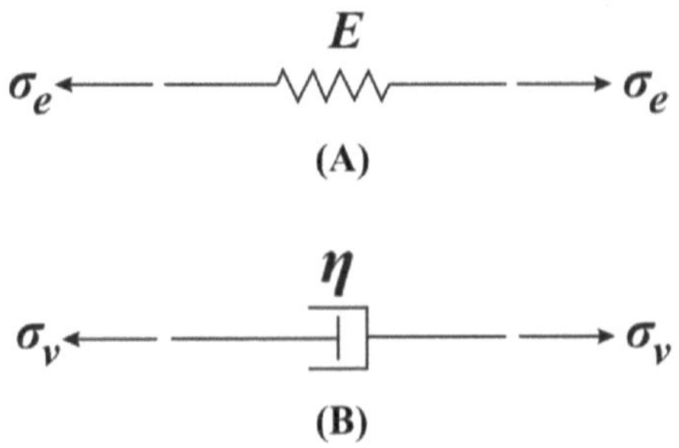

Figure 3.1 Components of integer-order viscoelastic models including (A) linear spring and (B) dashpot.

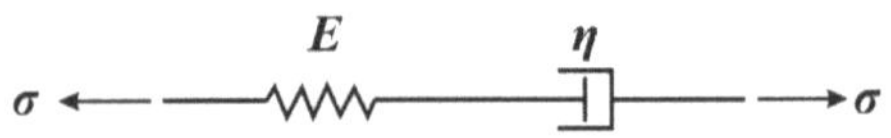

Figure 3.2 The Maxwell model.

3.2.1 The Maxwell model

The Maxwell model, which was established by Maxwell (1867), is constructed by connecting a linear spring and a dashpot in series, as shown in Fig. 3.2. The total stress and strain of the Maxwell model can be determined as indicated in Eq. (3.3):

$$\begin{aligned} &\sigma = \sigma_e = \sigma_v \\ &\varepsilon = \varepsilon_e + \varepsilon_v \Rightarrow \dot{\varepsilon} = \dot{\varepsilon}_e + \dot{\varepsilon}_v \end{aligned} \tag{3.3}$$

As can be seen in Eq. (3.3), it is possible to write the total strain of the Maxwell model in the first derivative format. Substituting Eqs. (3.1) and (3.2) into Eq. (3.3) gives the following equation for the total strain of the Maxwell model based on its components' coefficients:

$$\dot{\varepsilon} = \frac{\dot{\sigma}}{E} + \frac{\sigma}{\eta} \tag{3.4}$$

In order to determine constitutive functions of the Maxwell model, including creep compliance and relaxation modulus, a creep test under the applied stress of σ_0 and a relaxation test under the applied strain of ε_0

is assumed. Eq. (3.4) can be represented and integrated for $\sigma = \sigma_0$ as follows:

$$\begin{aligned} &\frac{d\varepsilon}{dt} = \frac{1}{E}\frac{d\sigma}{dt} + \frac{\sigma}{\eta} \Rightarrow d\varepsilon = E d\sigma + \frac{\sigma}{\eta} dt \\ &\int_0^t d\varepsilon = \int_0^t \frac{1}{E} d\sigma + \int_0^t \frac{\sigma}{\eta} dt \xrightarrow{\sigma(t) = \sigma_0} \varepsilon(t) = \left(\frac{1}{E} + \frac{t}{\eta}\right)\sigma_0 \end{aligned} \tag{3.5}$$

Implementing the definition of creep compliance (Eq. (2.1)), it is simply concluded that the creep compliance of the Maxwell model is as Eq. (3.6), which is linearly dependent on time. Fig. 3.3 depicts the response strain of the Maxwell model under a creep excitation and after removing the load. As can be seen in this figure, there is an instantaneous strain due to linear spring followed by linear incremental strain due to dashpot. After removing the load (unloading), the strain due to linear spring fully recovered instantaneously but the viscous portion of the strain cannot recover. Therefore the Maxwell model cannot reflect the recovery of viscoelastic materials and just represents elastic recovery.

$$D(t) = \frac{\varepsilon(t)}{\sigma_0} = \frac{1}{E} + \frac{t}{\eta} \tag{3.6}$$

For the relaxation test ($\varepsilon = \varepsilon_0$), $d\varepsilon/dt = 0$. Thus Eq. (3.4) can be considered as a first-order differential equation to determine relaxed stress during the relaxation test as represented in Eq. (3.7). It is necessary to mention that at the beginning of the test ($t = t_0$) for a sudden application of the strain, the dashpot behaves like a rigid element and the linear spring endures all applied strain. Then, the dashpot starts to work and the response stress gradually decreases.

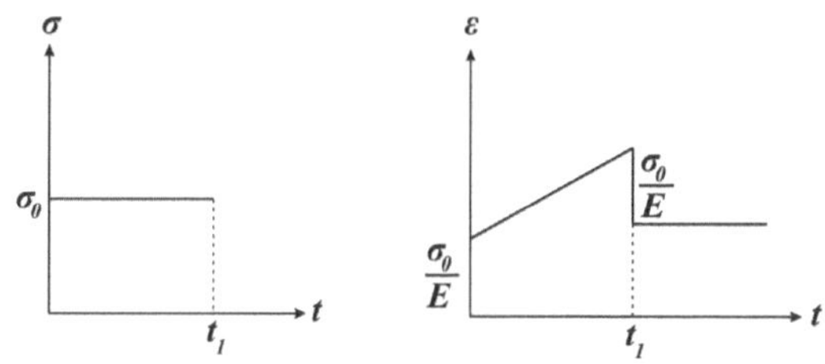

Figure 3.3 Creep behavior of the Maxwell model: during the loading and after unloading.

$$\frac{1}{E}\frac{d\sigma}{dt}+\frac{\sigma}{\eta}=0 \Rightarrow \sigma(t)=\exp\left[\int_t \frac{E}{\eta}dt\right]=Ce^{-\frac{E}{\eta}t}$$
$$t=0 \Rightarrow \sigma=E\varepsilon_0 \Rightarrow C=E\varepsilon_0 \tag{3.7}$$

Considering the Eq. (2.4) for a relaxation test, the relaxation modulus of the Maxwell model is an exponential function indicated in Eq. (3.8). The ratio of viscosity coefficient of the dashpot to the stiffness of the linear spring ($\tau=\eta/E$) is called relaxation time. The relaxation time for the Maxwell model is the time needed for the stress to relax and reduce to 36.8% of the initial value ($\sigma_0=E\varepsilon_0$).

$$E(t)=\frac{\sigma(t)}{\varepsilon_0}=E\exp\left(-\frac{E}{\eta}t\right) \tag{3.8}$$

3.2.2 The Kelvin–Voigt model

This model was established by Kelvin (1875) and Voigt (Voigt, 1892), hence being called the Kelvin–Voigt model. It is constructed by parallel connection of a linear spring and a dashpot, as shown in Fig. 3.4. The total stress and strain of the Kelvin–Voigt model can be evaluated as presented in Eq. (3.9):

$$\begin{aligned}\varepsilon&=\varepsilon_e=\varepsilon_v\\ \sigma&=\sigma_e+\sigma_v\end{aligned} \tag{3.9}$$

Substituting Eqs. (3.1) and (3.2) into Eq. (3.9) gives the following equation for total stress of the Kelvin–Voigt model based on its components' coefficients:

$$\sigma=E\varepsilon_e+\eta\dot{\varepsilon}_v \xrightarrow{\varepsilon=\varepsilon_e=\varepsilon_v} \sigma=E\varepsilon+\eta\dot{\varepsilon} \tag{3.10}$$

To determine constitutive functions of the Kelvin–Voigt model, including relaxation modulus and creep compliance, a relaxation test

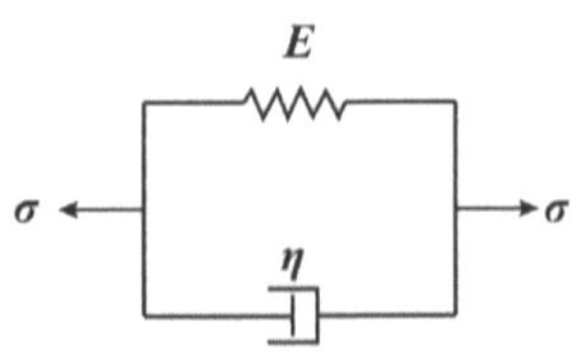

Figure 3.4 The Kelvin–Voigt model.

under the applied strain of ε_0 and a creep test under the applied stress of σ_0 is assumed. Eq. (3.10) is a first-order differential equation that can be solved by considering $\sigma = \sigma_0$ to determine creep compliance of the Kelvin–Voigt model. As can be seen in Eq. (3.11), the initial value ($\varepsilon = 0$ @ $t = 0$) used for solving this differential equation comes from the fact that at the beginning of the loading, when a load is suddenly applied, the dashpot reacts as a rigid element. The ratio of viscosity coefficient of the dashpot to stiffness of the linear spring ($\rho = \eta/E$) is called retardation time for the Kelvin–Voigt model. The retardation time is the time required for the strain to reach 63.2% of the total retarded strain.

$$\begin{aligned} &\dot{\varepsilon} + \frac{E}{\eta}\varepsilon = \frac{\sigma_0}{\eta} \Rightarrow \varepsilon(t) = \frac{\sigma_0}{E} + C\exp\left(-\frac{E}{\eta}t\right) \\ &t = 0 \Rightarrow \varepsilon = 0 \Rightarrow \varepsilon(t) = \frac{\sigma_0}{E}\left[1 - \exp\left(-\frac{E}{\eta}t\right)\right] \end{aligned} \tag{3.11}$$

According to the definition of creep compliance (Eq. (2.1)), the creep compliance of the Kelvin–Voigt model can be simply determined as Eq. (3.12). Fig. 3.5 shows the response strain of the Kelvin–Voigt model during a creep test and after unloading. After removing the load (unloading), the strain due to dashpot gradually recovered and tended to zero for infinite time. Therefore the Kelvin–Voigt model shows a full recovery.

$$D(t) = \frac{\varepsilon(t)}{\sigma_0} = \frac{1}{E}\left(1 - \exp\left[-\frac{E}{\eta}t\right]\right) \tag{3.12}$$

In a relaxation test when a sudden constant strain ($\varepsilon = \varepsilon_0$) should be applied on the Kelvin–Voigt model, it requires an instantaneous stress of

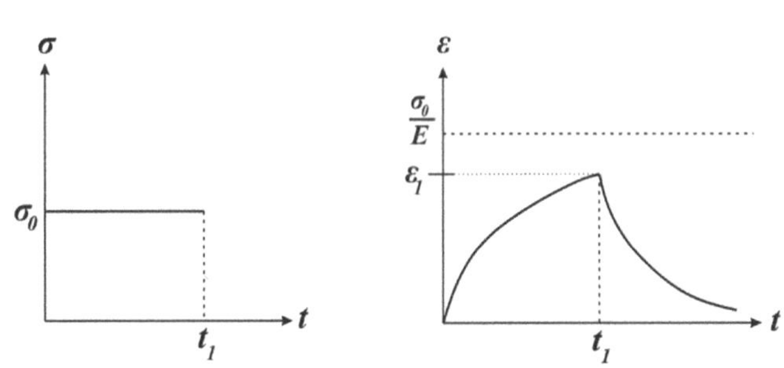

Figure 3.5 Creep behavior of the Maxwell model: during the loading and after unloading.

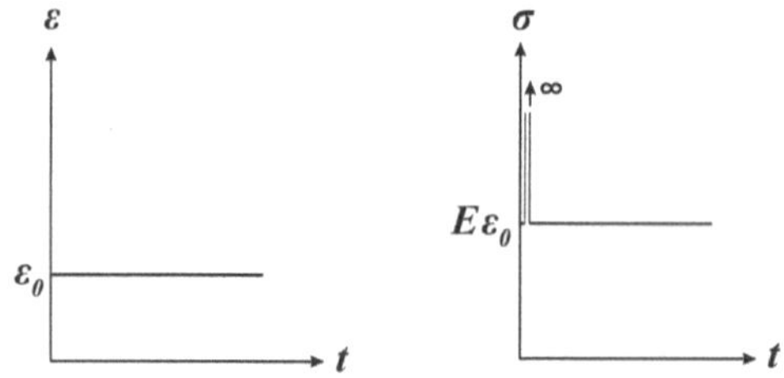

Figure 3.6 Relaxation behavior of the Kelvin–Voigt model.

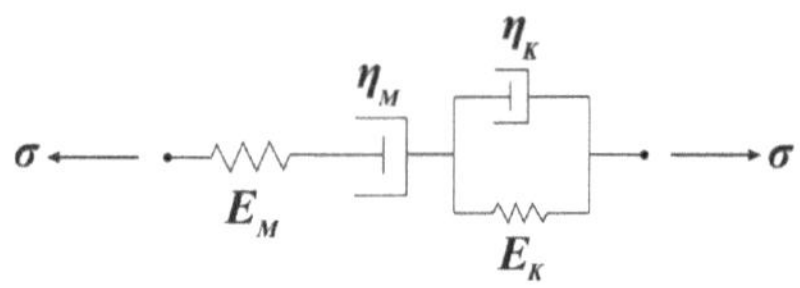

Figure 3.7 The Burgers model.

$\sigma = E\varepsilon_0$. As previously mentioned, the dashpot at sudden load application behaves like a rigid element. Therefore an infinitely large stress is needed to deform the Kelvin–Voigt model and reach a steady-state strain of ε_0. After that the stress will take the magnitude of $\sigma = E\varepsilon_0$. Fig. 3.6 depicts a relaxation test and correspondent stress for the Kelvin–Voigt model.

3.2.3 The Burgers model

As shown in Figs. 3.3 and 3.5, both the Maxwell model and the Kelvin–Voigt have shortcomings for capturing the creep-recovery behavior of viscoelastic materials. The Maxwell model can just show elastic recovery and the Kelvin–Voigt model shows full recovery without the ability to capture permanent deformation of viscoelastic materials. To overcome these shortcomings, a simple idea is to combine the simple Maxwell and Kelvin models benefit from both of them. The Burgers model is constructed by connecting a Maxwell model and a Kelvin–Voigt model in series as shown in Fig. 3.7. Burgers (1935) suggested this model to represent a pretty realistic viscoelastic behavior.

Since the Maxwell and Kelvin–Voigt model are connected in series, the stress and strain of the Burgers model can be described as follows:

$$\begin{aligned} \sigma_B &= \sigma_M = \sigma_K \\ \varepsilon_B &= \varepsilon_M + \varepsilon_K \end{aligned} \tag{3.13}$$

where subscripts B, M, and K refer to Burgers, Maxwell, and Kelvin–Voigt models, respectively. For a creep test under a constant stress of σ_0, it is possible to insert Eq. (3.5) for the Maxwell model and Eq. (3.11) for the Kelvin–Voigt model into Eq. (3.13) to obtain the total strain of the Burgers model as:

$$\varepsilon(t) = \sigma_0 \left[\underbrace{\left(\frac{1}{E_M} + \frac{t}{\eta_M} \right)}_{Maxwell\ \ model} + \underbrace{\frac{1}{E_K} \left(1 - \exp\left(-\frac{E_K}{\eta_K} t \right) \right)}_{Kelvin-Voigt\ \ model} \right] \tag{3.14}$$

Regarding Eq. (3.14) and considering Eq. (2.1), the creep compliance of the Burgers model can be simply described as Eq. (3.15):

$$D(t) = \frac{1}{E_M} + \frac{t}{\eta_M} + \frac{1}{E_K}\left(1 - e^{-\frac{E_K}{\eta_K}t}\right) \tag{3.15}$$

Fig. 3.8 depicts the creep and recovery behavior of the Burgers model. As can be seen in this figure, both instantaneous elastic recovery and

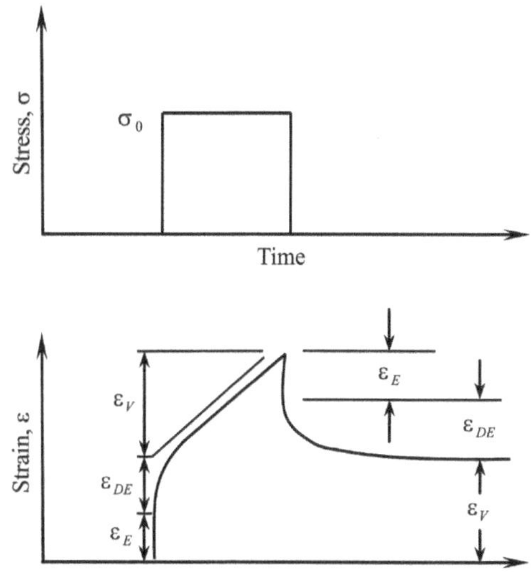

Figure 3.8 Creep and recovery behavior of the Burgers model. *From Aflaki, S., & Hajikarimi, P. (2012). Implementing viscoelastic rheological methods to evaluate low temperature performance of modified asphalt binders.* Construction and Building Materials, 36, *110–118. https://doi.org/10.1016/j.conbuildmat.2012.04.076.*

permanent deformation due to individual dashpot of the Maxwell model are captured by using the Burgers model. Regarding Fig. 3.8, after removing the load, elastic strain (ε_E) returns immediately, delayed elastic strain (ε_{DE}) recovers by passing a significant time, while viscous strain (ε_V) remains in the viscoelastic material as a permanent deformation and cannot recover.

It is possible to derive a differential equation of the Burgers model by implementing Eq. (3.13) as well as fundamental relations of the Maxwell and Kelvin–Voigt models. Considering Eq. (3.10) and knowing that $\sigma = \sigma_M = \sigma_K$, the stress of the Burgers model can be explained as follows:

$$\sigma = E_K(\varepsilon - \varepsilon_M) + \eta_K(\dot{\varepsilon} - \dot{\varepsilon}_M) \tag{3.16}$$

By differentiating Eq. (3.16) with respect to time and substituting Eq. (3.4) for the Maxwell model into this equation, the following second-order differential equation can be derived, which shows the stress–strain relation of the Burgers model (Ottosen & Ristinmaa, 2005):

$$\frac{\eta_K}{E_M}\ddot{\sigma} + \left(1 + \frac{\eta_K}{\eta_M} + \frac{E_K}{E_M}\right)\dot{\sigma} + \frac{E_K}{\eta_M}\sigma = \eta_K\ddot{\varepsilon} + E_K\dot{\varepsilon} \tag{3.17}$$

Aflaki and Hajikarimi (2012) implemented the Burgers model to investigate the effect of different types of additive on low-temperature creep behavior of asphalt binders using the bending beam rheometer (BBR) test (described in Section 2.1). For this purpose, they derived stored and dissipated energy in terms of the Burgers model parameters as follows:

$$W_s(t) = \sigma_0^2\left[\frac{1}{E_M} + \frac{1}{2E_K}\left(1 - 2e^{-\frac{E_M}{\eta_M}t} + e^{-2\frac{E_K}{\eta_K}t}\right)\right] \tag{3.18}$$

$$W_d(t) = \sigma_0^2\left[\frac{t}{\eta_0} + \frac{1}{2E_M}\left(1 - e^{-2\frac{E_K}{\eta_K}t}\right)\right] \tag{3.19}$$

where $W_s(t)$ is stored energy, $W_d(t)$ is dissipated energy, t is time of loading, and σ_0 can be calculated as:

$$\sigma_0 = \frac{3PL}{2bh^2} \tag{3.20}$$

which is 0.293 MPa under the BBR test conditions. Then, having both stored and dissipated energy, the dissipated energy ratio $[W_d(t)/W_s(t)]$ was defined when t tends to infinity ($t \rightarrow \infty$). The stored and dissipated energy for $t \rightarrow \infty$ can be determined as:

$$W_s(t \rightarrow \infty) = \sigma_0^2 \left(\frac{1}{E_M} + \frac{1}{2E_K} \right) \tag{3.21}$$

$$W_d(t \rightarrow \infty) = \sigma_0^2 \left(\frac{t}{\eta_M} + \frac{1}{2E_1} \right) \tag{3.22}$$

For limiting the value of dissipated energy, a specific time of $t = 60$ s was selected in agreement with ASTM D 6648. Fig. 3.9 depicts dissipated energy ratio for asphalt binders modified with crumb rubber (Cr with 10%, 14%, and 16%), gilsonite (G with 2%, 4%, 7%, 10%, and 13%), styrene-butadiene-styrene (S with 1%, 2%, and 3%), and polyphosphoric acid (P with 0.5%, 0.7%, 1.3%, and 1.6%) at different test temperatures of −6°C, −12°C, −18°C, −24°C, and −30°C. By decreasing the test temperature, the dissipated energy ratio decreases, which is expected due to increasing the asphalt binder's stiffness. Generally, a higher dissipated energy ratio reflects better low-temperature behavior of asphalt binders.

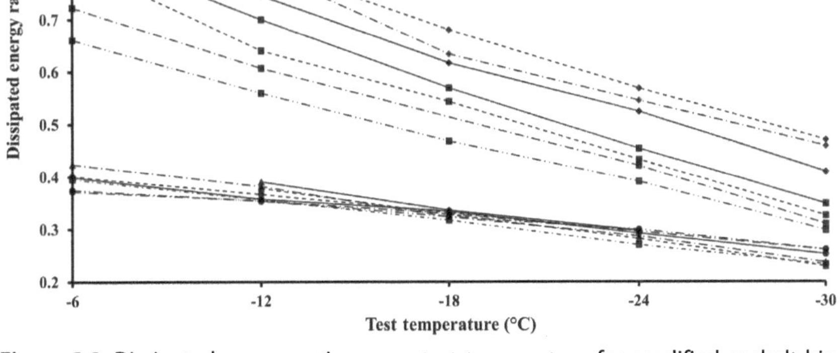

Figure 3.9 Dissipated energy ratio versus test temperature for modified asphalt binders at loading time of 60 s. *From Aflaki, S., & Hajikarimi, P. (2012). Implementing viscoelastic rheological methods to evaluate low temperature performance of modified asphalt binders.* Construction and Building Materials, 36, *110–118. https://doi.org/10.1016/j.conbuildmat.2012.04.076.*

To evaluate high-temperature performance of original and modified asphalt binders, Bahia et al. introduced a new parameter G_v defined as the viscous component of creep stiffness of the Burgers model based on the repeated creep and recovery (RCR) test results as shown in Eq. (3.23).

$$G_v = \frac{1}{J_v} = \frac{\eta_M}{t} \tag{3.23}$$

where J_v is the viscous creep compliance related to the dashpot of the Maxwell model. This parameter can reflect the permanent deformation of asphalt binders at high temperatures. A schematic result of a RCR test for one cycle is shown in Fig. 3.10 in which a shear stress in the range of 30–300 Pa for 100 cycles with 1 s loading time followed by a 9 s unloading time was suggested based on the NCHRP 9–10 project. As the viscous component of creep stiffness increases, the dissipated energy decreases and antirutting potential also reduces.

3.2.4 The generalized Maxwell or Kelvin–Voigt model

In order to improve the accuracy of predicting creep compliance or relaxation modulus of viscoelastic materials in a broader range of frequency or time, a brilliant idea is to combine several Kelvin–Voigt models in series or several Maxwell models in parallel as depicted in Figs. 3.11 and 3.12, respectively.

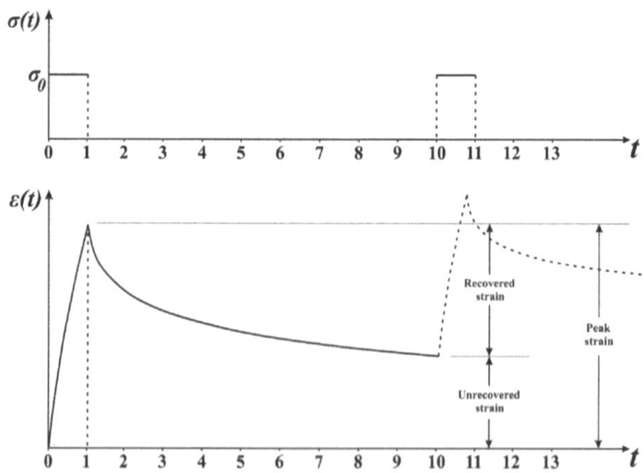

Figure 3.10 Schematic loading, un-loading, and results of the repeated creep and recovery test for one cycle (Hajikarimi, Rahi, & Moghadas Nejad, 2015).

The response of the generalized Kelvin–Voigt model in a creep test can be determined by using Eq. (3.11) and adding n terms as follows:

$$D(t) = D_g + \sum_{i=1}^{n} D_i\left(1 - \exp\left[-\frac{t}{\rho_i}\right]\right) \tag{3.24}$$

where $D(t)$ is creep compliance, D_g is glassy compliance, D_i is compliance of each spring ($1/E_i$), and ρ_i is retardation time. The response of the generalized Maxwell model in a relaxation test can be described referring to Eq. (3.8) by adding n terms as:

$$E = E_\infty + \sum_{i=1}^{n} E_i \exp\left(-\frac{t}{\tau_i}\right) \tag{3.25}$$

where $E(t)$ is relaxation modulus, E_∞ is equilibrium modulus, E_i is stiffness of each spring, and τ_i is relaxation time. The equation mentioned above for relaxation modulus of a generalized Maxwell model is known as Prony series. For using Prony series to describe the mechanical behavior of a viscoelastic material, the question then arises about the selection of n

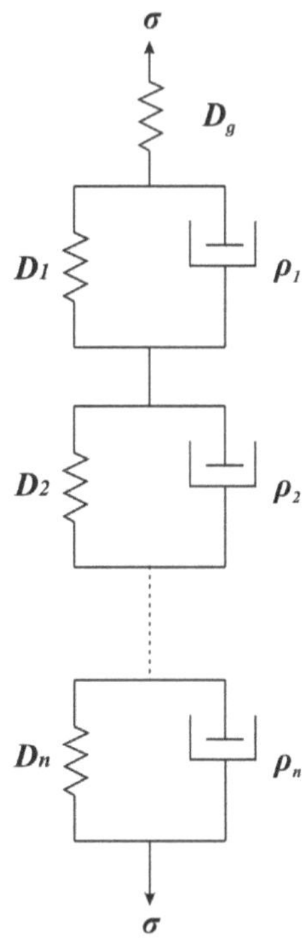

Figure 3.11 The generalized Kelvin–Voigt model.

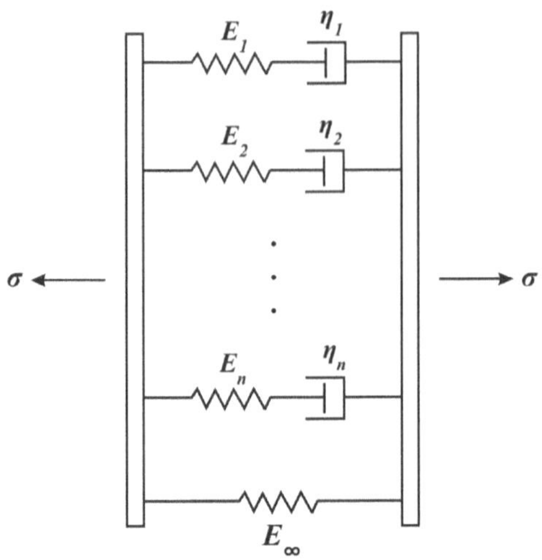

Figure 3.12 The generalized Maxwell model.

to be adequate for accurately reflecting relaxation behavior. For this purpose, Hajikarimi et al. (2018b) constructed several generalized Maxwell models (considering different number of n) and hypothesized the statistical Fisher test (F-test) with a confidence level of $\alpha = 0.05$ on each model. The F-test is a common statistical method to compare nested models. Two models are nested if both contain the same terms and one has at least an additional term. Consequently, a generalized Maxwell model with n arms (model-A) and another one with $n+1$ arms (model-B) are nested models. These models are needed to evaluate whether a more complex model can provide any additional information. In their approach, the F-test consists of comparisons between the sum of squares of errors (SSE) for model-A (SSE_A) and model-B (SSE_B). Generally, it is evident that always $SSE_A > SSE_B$, so the question then arises whether the model-B reduction of SSE is large enough. Thus based on the F-test concept, the hypothesis can be explained as:

$$H_0{:}E_{n+1} = \tau_{m+1} = 0 \ \textit{versus} \ H_\alpha{:}E_{n+1} \neq 0 \ \textit{and} \ \tau_{n+1} \neq 0 \tag{3.26}$$

in which the F-test statistic is defined as:

$$F = \frac{(SSE_A - SSE_B)/(\textit{number of additional parameters} = 2)}{SSE_B/[m - 2(n+1)]} \tag{3.27}$$

where m is the number of self-governing data points and n is the number of Maxwell components in model-A which is assumed incomplete. At a confidence level of α, the F-statistic is compared with an $F_{2,\ [m-2(n+1)]}$ based on the F-test table. If $F \geq F_{\alpha,2,\ [m-2(n+1)]}$, H_0 is rejected. Fig. 3.13 shows generalized Maxwell models with different number of n (2, 5, and 10 arms) fitted on an experimental master curve of an asphalt binder. Table 3.1 also shows the associated results of F-test for this sample with $\alpha = 0.05$, which is a common and satisfactory consideration. According to these outputs, it is evident that using nine Maxwell models in parallel and an individual spring (19 parameters including E_i and τ_i) passes the test and can sufficiently represent the relaxation behavior of the asphalt binders under creep condition of BBR test at low test temperatures.

A significant issue for using generalized Maxwell or Kelvin–Voigt models is finding coefficients of Eq. (3.24) or Eq. (3.25). For this purpose, a simple MATLAB code is presented in Box 3.1 that can be used for determining these coefficients. As can be seen in this figure, using the "fit" function of MATLAB software and defining some conditions for D_g, D_i, and ρ_i as well as desired R squared, it is well-known that the curve-fitting toolbox of MATLAB software can also be used by calling CFTOOL in command window and then defining a costume function to find the required coefficients based on the nonlinear regression method.

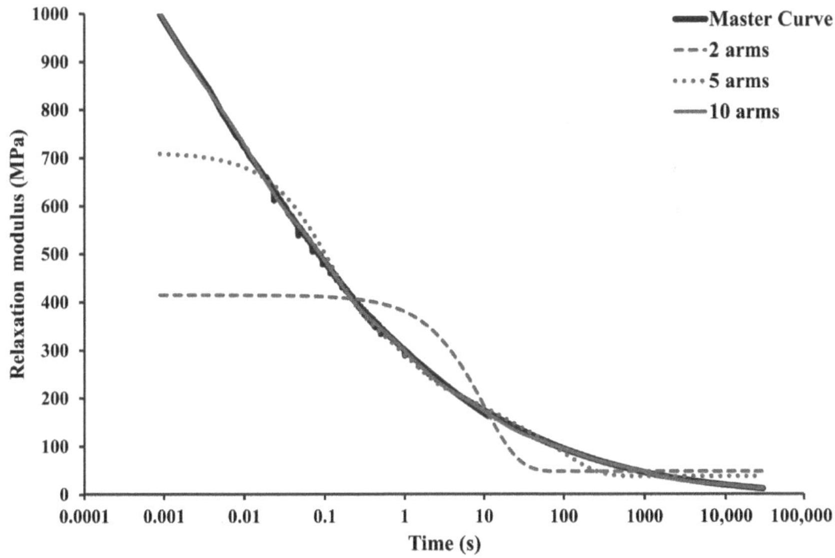

Figure 3.13 Fitting generalized Maxwell models on the relaxation modulus data with different numbers of Maxwell models.

Table 3.1 Results of the F-test for an asphalt binder sample.

$n+1$	F	$F_{a,2[m-2(n+1)]}$	F-test result
3	888.4	2.9995	Failed
4	1767.44	2.9995	Failed
5	1921.14	2.9995	Failed
6	1149.1	2.9995	Failed
7	9315.22	2.9995	Failed
8	284.58	2.9995	Failed
9	1205	2.9995	Failed
10	9.36	2.9995	Failed
11	0	2.9995	Passed

BOX 3.1 A MATLAB code for determining creep compliance coefficients of the generalized Kelvin–Voigt model.

Creep_Compliance.m

```
%Determination of Creep Compliance Function Coefficients
clc
clear f D p1 Coeffs R2 D_g D_i Tau_i

while 1>0 %Curve Fitting on (D(t) versus t) ==> Conditions: R2>0.999 & D_i>0 & Tau_i>0
  [f,p1]=fit(t,D,'D0+D1*(1-exp(-x/Tau1))+D2*(1-exp(-x/Tau2))+...+D9*(1-exp(-x/Tau9))',...
      'lower',[0,0,0,0,0,0,0,0,0,0,0,0,0,0,0,0,0,0],...
      'upper',[100,100,100,100,100,100,100,100,100,100,100,100,100,100,100,100,100,100]);
  Coeffs = coeffvalues(f);
  R2 = p1.rsquare;
  if Coeffs(1)>0 && Coeffs(2)>0 && Coeffs(3)>0 && Coeffs(4)>0 && Coeffs(5)>0 && Coeffs(6)>0 && Coeffs(7)>0 && ...
          Coeffs(8)>0 && Coeffs(9)>0 && Coeffs(10)>0 && Coeffs(11)>0 && Coeffs(12)>0 && Coeffs(13)>0 && ...
          Coeffs(14)>0 && Coeffs(15)>0 && Coeffs(16)>0 && Coeffs(17)>0 && Coeffs(18)>0 && Coeffs(19)>0 && R2>0.999
      break
  end
end

D_0 = Coeffs(1);
D_i = [Coeffs(2) Coeffs(3) Coeffs(4) Coeffs(5) Coeffs(6) Coeffs(7) Coeffs(8) Coeffs(9) Coeffs(10)];
Tau_i = [Coeffs(11) Coeffs(12) Coeffs(13) Coeffs(14) Coeffs(15) Coeffs(16) Coeffs(17) Coeffs(18) Coeffs(19)];
```

3.2.5 General format of the constitutive equation

The previous sections have mainly focused on deriving functions of creep compliance and relaxation modulus for different configurations of springs and dashpots by considering a special case of creep or relaxation test. In addition to constitutive viscoelastic functions to characterize viscoelastic materials, a general stress–strain relation may be necessary, especially for use in numerical simulations. The general format of the constitutive integer-order equation is as follows:

$$\sum_{i=1}^{n} a_i \frac{d^i \sigma}{dt^i} = \sum_{j=1}^{m} b_j \frac{d^j \varepsilon}{dt^j} \tag{3.28}$$

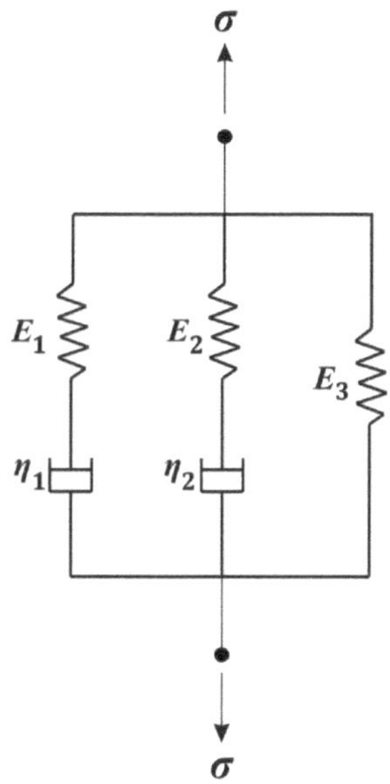

Figure 3.14 A spring-dashpot model for solid viscoelastic materials.

in which d^i/dt^i is the derivative operation, a_i and b_j are constant material parameters, and generally m is not equal to n. Eq. (3.17) shows such an equation for explaining the stress–strain relation of the Burgers model. The main drawback of using integer-order differential equations for explaining the stress–strain relation of viscoelastic materials can be seen as it becomes cumbersome for a more advanced model. For better illustrating this drawback, a generalized Maxwell model with two arms and a single spring is considered, as can be seen in Fig. 3.14. The following equations can be simply written for this model:

$$\begin{aligned} &\sigma = \sigma_1 = \sigma_2 = \sigma_3 \Rightarrow \dot{\sigma} = \dot{\sigma}_1 = \dot{\sigma}_2 = \dot{\sigma}_3 \\ &\varepsilon = \varepsilon_1 = \varepsilon_2 = \varepsilon_3 \end{aligned} \tag{3.29}$$

Considering Eq. (3.1) for the single spring as well as Eq. (3.4) for Maxwell models, it is possible to write the following equations:

$$\begin{aligned} \dot{\varepsilon}_1 &= \frac{\dot{\sigma}_1}{E_1} + \frac{\sigma_1}{\eta_1} \\ \dot{\varepsilon}_2 &= \frac{\dot{\sigma}_2}{E_2} + \frac{\sigma_2}{\eta_2} \\ \dot{\varepsilon}_3 &= \frac{\dot{\sigma}_3}{E_3} \end{aligned} \tag{3.30}$$

By using Eq. (3.29), it is possible to derive the following equation:

$$\sigma = \sigma_1 + \sigma_2 + \sigma_3 \xrightarrow{/\eta_1} \frac{\sigma}{\eta_1} = \frac{\sigma_1}{\eta_1} + \frac{\sigma_2}{\eta_1} + \frac{\sigma_3}{\eta_1}$$

$$\dot{\sigma} = \dot{\sigma}_1 + \dot{\sigma}_2 + \dot{\sigma}_3 \xrightarrow{/E_1} \frac{\dot{\sigma}}{E_1} = \frac{\dot{\sigma}_1}{E_1} + \frac{\dot{\sigma}_2}{E_1} + \frac{\dot{\sigma}_3}{E_1}$$

$$\dot{\sigma} + \sigma = (\dot{\sigma}_1 + \sigma_1) + (\dot{\sigma}_2 + \sigma_2) + (\dot{\sigma}_3 + \sigma_3)$$

$$= \underbrace{\left[\frac{\dot{\sigma}_1}{E_1} + \frac{\sigma_1}{\eta_1}\right]}_{\dot{\varepsilon}_1 = \dot{\varepsilon}} + \left[\frac{\dot{\sigma}_2}{E_1} + \frac{\sigma_2}{\eta_1}\right] + \left[\frac{\dot{\sigma}_3}{E_1} + \frac{\sigma_3}{\eta_1}\right] \tag{3.31}$$

$$\Rightarrow \frac{\dot{\sigma}}{E_1} + \frac{\sigma}{\eta_1} = \dot{\varepsilon} + \frac{\dot{\sigma}_2}{E_1} + \frac{\sigma_2}{\eta_1} + \frac{E_3\dot{\varepsilon}}{E_1} + \frac{E_3\varepsilon}{\eta_1}$$

$$\Rightarrow \frac{\dot{\sigma}}{E_1} + \frac{\sigma}{\eta_1} = \frac{(E_1 + E_2 + E_3)}{E_1}\dot{\varepsilon} + \frac{E_3}{\eta_1}\varepsilon + \sigma_2\left(\frac{E_1\eta_2 - E_2\eta_1}{E_1\eta_1\eta_2}\right)$$

By differentiating Eq. (3.31) with respect to time and dividing both sides of the equation by E_2, the following equation can be derived:

$$\frac{\ddot{\sigma}}{E_1E_2} + \frac{\dot{\sigma}}{\eta_1E_2} = \frac{E_3}{\eta_1E_2}\dot{\varepsilon} + \left(\frac{E_1 + E_2 + E_3}{E_1E_2}\right)\ddot{\varepsilon} + \left(\frac{E_1\eta_2 - E_2\eta_1}{E_1E_2\eta_1\eta_2}\right)\dot{\sigma}_2 \tag{3.32}$$

Then, Eq. (3.31) which is divided by η_2 and Eq. (3.32) are added together as:

$$\frac{\ddot{\sigma}}{E_1E_2} + \left(\frac{1}{\eta_1E_2} + \frac{1}{E_1\eta_2}\right)\dot{\sigma} + \frac{1}{\eta_1\eta_2}\sigma = \left(\frac{E_1 + E_2 + E_3}{E_1E_2}\right)\ddot{\varepsilon}$$
$$+ \left(\frac{E_3}{E_2\eta_1} + \frac{E_1 + E_2 + E_3}{E_1\eta_2} + \frac{\eta_2 - \eta_1}{\eta_1\eta_2}\right)\dot{\varepsilon} + \frac{E_3}{\eta_1\eta_2}\varepsilon \tag{3.33}$$

which is a second-order differential equation for describing the general stress–strain relation of the parallel connection of two Maxwell models with an individual spring, as depicted in Fig. 3.14. It is apparent that increasing the number of springs and dashpots resulted in increasing the order of the corresponding differential equations which causes the equation derivation procedure to be more difficult and cumbersome.

3.3 Fractional-order differential equation

The integer-order viscoelastic models have the advantage of using spring-dashpot components to describe linear behavior viscoelastic materials. As mentioned in Section 3.2, the accurate description of the viscoelastic response with such models normally needs a series of model components. For example, the generalized Maxwell model requires 8–25 elements to describe the mechanical behavior of a bituminous material accurately. Therefore using integer-order models means that many parameters need to be determined to form the desired model. The large number of parameters leads to increasing the calculation time. Furthermore, a model with several parameters is not an appropriate alternative to be popularized and applied in practical cases. Thus the use of advanced models has been considered to decrease the number of parameters. Several fractional-order models have been developed to solve the aforementioned problem and decrease the required number of parameters without reducing the precision of the model. In these models, fractional components are included, such as parabolic dashpots. The Huet model was among the first models proposed in the literature (Huet, 1965). It is composed of two parabolic dashpots in series with a spring. Later, an improved version of this model called Huet–Sayegh (HS) model was developed by adding an elastic spring in parallel with the Huet one (Sayegh, 1967). The accurate description of viscoelastic mechanical behavior, based on the HS model, has been verified in many studies. Graziani et al. (2020) used the HS model to simulate the rheological experimental results of cold-recycled mixes constructed in-place by full-depth reclamation. It was found that the HS model affords an excellent fitting on the dynamic modulus, while fitting on the tangent loss angle was improved, using a temperature- and frequency-independent modification. Similar conclusions were obtained by Xu and Solaimanian (2009). Later, a new fractional viscoelastic model was proposed by Olard and Di Benedetto (2003) which is called 2S2P1D. This model contains one linear dashpot, two parabolic dashpots, and one elastic spring, which are in parallel with another elastic spring. Mangiafico et al. (2019) evaluated the relative difference between experimental data (viscoelastic properties of both bitumen and mixture) and simulation results with the analogical 2S2P1D model. The obtained results highlighted the suitability of the 2S2P1D model to conduct multiscale modeling of the linear viscoelastic response of bituminous materials. Lastly, Woldekidan and his coworkers (Woldekidan et al., 2012) established a modified Huet–Sayegh (MHS) model by introducing a linear dashpot in series with the basic HS model. The results of this research work showed the accuracy and suitability of the proposed model to be used for finite element simulations. In the literature, other types of fractional models were considered that use a fractional dashpot instead of the actual linear dashpot in the integer-order model. A fractional dashpot can involve two parameters similar to a parabolic dashpot. Though, the constitutive equations of both models are different. Xu et al. (2019) investigated several configurations of fractional viscoelastic models in order to establish an appropriate model that can simulate the viscoelastic characteristics of asphalt binders as well as asphalt mastics.

In this section, an introduction to fractional-order differentiation and differential equations is presented at first and then some of the most commonly used fractional viscoelastic models are introduced, and their application considered for modeling the mechanical behavior of viscoelastic materials.

3.3.1 Fractional derivatives

The history of fractional derivatives goes back to 1695 when L'Hopital and Leibniz discussed the meaningfulness of defining d^n/dt^n for $n = 0.5$. In 1921 Nutting (1921) observed that stress relaxation in some materials can be modeled by using the general power law and Gemant (1936) reported that stiffness and damping characteristics of viscoelastic materials can be better fitted by using fractional powers of frequency. Caputo and Mainardi (1971) also found good agreement between experimental results and an analytical solution based on fractional derivatives for viscoelastic materials. Until the early 1980s the concept of fractional derivatives was just known as a curve-fitting method within the viscoelasticity theorem framework. Then, Bagley and Torvik (1983) presented a physical and mechanical definition of this concept. There are different definitions for fractional derivatives. The starting point of these definitions is first-order integer derivative which can be simply written as follows:

$$\frac{d^1 f(t)}{dt^1} = \lim_{\Delta t \to \infty} \frac{f(t) - f(t - \Delta t)}{\Delta t} \tag{3.34}$$

The higher order (nth) derivative is presented as:

$$\frac{d^n f(t)}{dt^n} = \lim_{\Delta t \to \infty} \left[(\Delta t)^{-n} \sum_{j=0}^{n} (-1)^j \binom{n}{j} f(t - j\Delta t) \right] \tag{3.35}$$

where,

$$\binom{n}{j} = \begin{cases} \dfrac{n!}{j!(n-j)!}, & 0 \leq j \leq n \\ 0, & 0 \leq n \leq j \end{cases} \tag{3.36}$$

By substituting Δt with t/N ($N = 1, 2, 3, \ldots$) in Eq. (3.35), this equation can be rewritten as:

$$\frac{d^n f(t)}{dt^n} = \lim_{N\to\infty} \left[\left(\frac{t}{N}\right)^{-n} \sum_{j=0}^{N-1} (-1)^n \binom{n}{j} f\left(t - j\frac{t}{N}\right) \right] \tag{3.37}$$

with respect to $(n,j) = 0$ for $j > n$. In order to derive an equation which is valid for any real-order derivative, the extension of Eq. (3.36) is presented as:

$$\binom{a}{j} = \begin{cases} \dfrac{a(a-1)(a-2)\ldots(a-j+1)}{j!}, & j > 0 \\ 1, & j = 0 \end{cases} \tag{3.38}$$

where a is a real number and j is a positive integer one. For $j > 0$, it is possible to write:

$$\begin{aligned} (-1)^j \binom{n}{j} &= (-1)^j \frac{n(n-1)(n-2)\ldots(n-j+2)(n-j+1)}{j!} \\ &= \frac{(j-n-1)(j-n-2)\ldots(-n+1)(-n)}{j!} \end{aligned} \tag{3.39}$$

and then, it can be concluded:

$$(-1)^j \binom{n}{j} = \binom{j-n-1}{j} = \frac{\Gamma(j-n)}{\Gamma(-n)\Gamma(j+1)} \tag{3.40}$$

in which $\Gamma(.)$ is gamma function. By inserting Eq. (3.40) into Eq. (3.37):

$$\frac{d^n f(t)}{dt^n} = \lim_{N\to\infty} \left[\left(\frac{t}{N}\right)^{-n} \sum_{j=0}^{N-1} A_{j+1} f\left(t - j\frac{t}{N}\right) \right] \tag{3.41}$$

where

$$A_{j+1} = \frac{\Gamma(j-n)}{\Gamma(-n)\Gamma(j+1)} \tag{3.42}$$

which is valid for each real-order derivative ($n > 0$). By substituting the variable of n with a notation of α as an arbitrary real number, the Granwald definition can be derived for fractional derivatives as follows:

$$\frac{d^{\alpha}f(t)}{dt^{\alpha}} = \lim_{N\to\infty}\left[\left(\frac{t}{N}\right)^{-\alpha}\sum_{j=0}^{N-1}A_{j+1}f\left(t-j\frac{t}{N}\right)\right] \tag{3.43}$$

in which

$$A_{j+1} = \frac{\Gamma(j-\alpha)}{\Gamma(-\alpha)\Gamma(j+1)} \tag{3.44}$$

It is necessary to mention that all Granwald A_{j+1} coefficients are non-zero if the order of derivation, α, is not a positive integer figure. For example, if $\alpha = -1$ then for all j, $A_{j+1} = 1$.

The lower limit of the fractional derivation represented in Eq. (3.43) is zero. The first term of this series for $j = 0$ is $A_1 f(t)$ and the last term for $j = N$ is $A_N f(t/N)$. Therefore the range of $(0, t]$ is divided into N equal intervals which are used to calculate the fractional derivatives or fractional integrals. From a mechanical point of view, the lower limit of the fractional derivative represents the moment from which the history of $f(t)$ is considered, which is similar to ordinary integrals. The modified definition of Granwald which is based on central differences is as follows:

$$\frac{d^{\alpha}f(t)}{dt^{\alpha}} = \lim_{N\to\infty}\left[\left(\frac{t}{N}\right)^{-\alpha}\sum_{j=0}^{N-1}A_{j+1}f\left(t-\left(j-\frac{\alpha}{2}\right)\frac{t}{N}\right)\right] \tag{3.45}$$

which converges to the original value faster and hence it is preferred to use this form of Granwald equation in numerical algorithms.

3.3.2 Simple fractional element

The ideal viscous behavior of liquids can be described by using a dashpot that obeys the following first-order differential equation (in terms of time):

$$\tau(t) = \eta\left[\frac{d}{dt}\gamma(t)\right] \tag{3.46}$$

where τ is stress, γ is strain, t is time, and η is viscosity coefficient of the viscous liquid. For a constant applied stress, the creep function of a dashpot can be written as follows:

$$\gamma(t) = \frac{t}{\eta}\tau_0 \tag{3.47}$$

It is possible to show the creep function of a dashpot schematically as it is represented in curve 1 of Fig. 3.15. As expected based on Eq. (3.47), the creep function behaves linearly during the loading time and after removing the load, the strain remains constant. Therefore a dashpot cannot release the strain which is absorbed during the loading time. In Fig. 3.15 curve 2 shows the strain curve of a linear spring in which the strain immediately develops after loading and remains constant for a constant applied stress. After removing the load, the strain dissipates instantaneously and the spring returns to its preliminary state. Since most of the viscous materials consist of solid and liquid components, a combination of springs and dashpots can well reflect its creep behavior. The fractional differential time equation (FDTE) can be used for this purpose as follows:

$$\tau(t) = E\frac{d^\alpha}{dt^\alpha}\gamma(t) \tag{3.48}$$

where α is the order of differential equation, p is a constant parameter, t is time, and τ and γ are stress and strain, respectively. Considering $\alpha = 0$, $\tau = p\ \gamma$ which shows the stress–strain behavior of linear elastic solids. On the other hand, by considering $\alpha = 1$, Eq. (3.48) will be same as Eq. (3.46) which shows the stress–strain behavior of ideal viscous liquids.

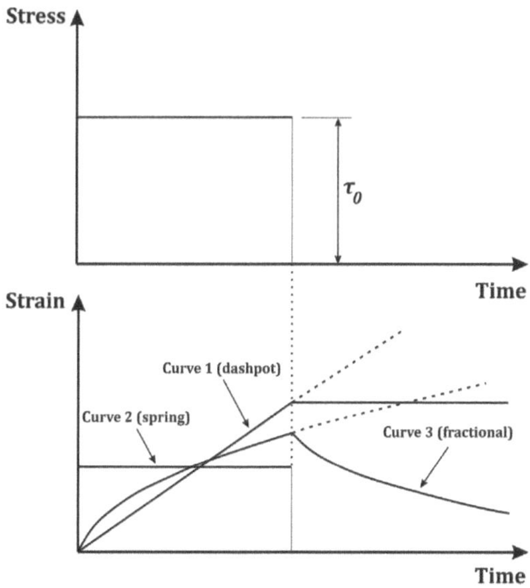

Figure 3.15 Creep function curves for a linear spring, a dashpot, and a fractional element.

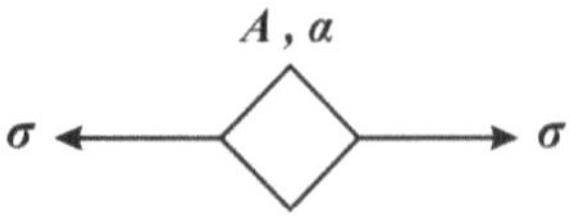

Figure 3.16 The simple fractional element.

However, by assuming $0 < \alpha < 1$, the correspondent creep strain function can be derived by implementing fractional integration as follows:

$$\gamma(t) = \frac{t^{\alpha}}{E\Gamma(\alpha+1)}\tau_0 \tag{3.49}$$

where Γ denotes the Gamma function. Curve 3 in Fig. 3.15 shows the nonlinear creep behavior of the fractional element which is capable of gradually dissipating the whole strain after removing the load. By using the definition of creep compliance function (Eq. (2.1)), it simply can be shown that:

$$D(t) = \frac{1}{E}\frac{t^{\alpha}}{\Gamma(\alpha+1)} = At^{\alpha} \tag{3.50}$$

where $D(t)$ is the creep compliance function and A is an equivalent constant parameter. The relaxation modulus also can be derived as follows:

$$E(t) = \frac{t^{-\alpha}}{A\Gamma(1+\alpha)\Gamma(1-\alpha)} \tag{3.51}$$

where A is the equivalent constant parameter introduced in Eq. (3.50). Having defined the relaxation modulus and creep compliance by using the same parameters, the simple fractional model can be used as a practical viscoelastic component, as depicted in Fig. 3.16.

By using the direct relation between storage modulus and loss modulus with relaxation modulus (Gutierrez-Lemini, 2014) (Eqs. (2.24) and (2.25)) and substituting Eq. (3.51) into Eq. (2.24), it is possible to determine the storage modulus in the following complex form:

$$E'(\omega) = \frac{\omega}{A\Gamma(1-\alpha)\Gamma(1+\alpha)} \left\{ -\frac{[(-i\omega)^{\alpha}\Gamma(1-\alpha, -it\omega) + (i\omega)^{\alpha}\Gamma(1-\alpha, -it\omega)]}{2\omega} \right\} \tag{3.52}$$

in which $\Gamma(s, x)$ is the incomplete Gamma function and is defined as:

$$\Gamma(s, x) = \int_x^{\infty} t^{s-1} e^{-t} dt \tag{3.53}$$

By applying the integral boundaries, the following equation can be derived:

$$\begin{aligned} &if t\rightarrow \infty;\Gamma(1-\alpha,-it\omega)=\Gamma(1-\alpha,it\omega)=0\\ &if t=0;\Gamma(1-\alpha,0)=\Gamma(1-\alpha)\end{aligned} \tag{3.54}$$

Then, it can be shown that the storage modulus can be rewritten as follows:

$$E'(\omega)=\frac{\omega^{\alpha}}{\Gamma(1+\alpha)}\left\{\frac{[(-i)^{\alpha}+(i)^{\alpha}]}{2}\right\} \tag{3.55}$$

By considering the following general relation

$$e^{i\omega}=\cos(\omega)+i\sin(\omega) \tag{3.56}$$

and assuming $\omega=\pi/2$ or $\omega=-\pi/2$, it is evident that:

$$\begin{aligned} (i)^{\alpha}&=e^{\frac{i\pi\alpha}{2}}\\ (-i)^{\alpha}&=e^{-\frac{i\pi\alpha}{2}}\end{aligned} \tag{3.57}$$

and by implementing Eq. (3.57) within Eq. (3.55), the storage modulus can be derived as:

$$E'(\omega)=\frac{\omega^{\alpha}}{A\Gamma(1+\alpha)}\cos\left(\frac{\pi\alpha}{2}\right) \tag{3.58}$$

Following a similar procedure, the loss modulus, $E''(\omega)$ can be derived as:

$$E''(\omega)=\frac{\omega^{\alpha}}{A\Gamma(1+\alpha)}\sin\left(\frac{\pi\alpha}{2}\right) \tag{3.59}$$

Having the storage modulus and loss modulus, it is possible to determine the damping ratio as follows:

$$Damping\ Ratio=\frac{E''(\omega)}{E'(\omega)}=\tan\left(\frac{\pi\alpha}{2}\right) \tag{3.60}$$

It can be seen that the damping ratio of the simple fractional element is independent of frequency and just related to parameter α. According to the tangent function properties, the greater amount of α leads to a higher damping ratio.

Furthermore, stored energy (W_s) and dissipated energy (W_d) can be determined for the simple fractional element as follows (Brinson & Brinson, 2008):

$$W_s(t) = A\tau_0^2\left[t^\alpha - \frac{1}{2}(2t)^\alpha\right] \tag{3.61}$$

$$W_d(t) = \frac{1}{2}\tau_0^2 A(2t)^\alpha \tag{3.62}$$

The parameter of dissipated energy ratio (*DER*) is defined as the division of dissipated energy by stored energy as:

$$DER = \frac{W_d(t)}{W_s(t)} \tag{3.63}$$

By substituting Eqs. (3.61) and (3.62) into Eq. (3.63), the *DER* can be determined for the simple fractional element as:

$$DER = \frac{2^{\alpha-1}}{1-2^{\alpha-1}} \tag{3.64}$$

3.3.3 Generalized fractional viscoelastic model

By inspiring the idea of generalized Maxwell and Kelvin model (Tschoegl, 1989) and some research works for determining relaxation modulus and creep compliance as two constitutive viscoelastic functions (Park & Kim, 1999; Stimilli et al., 2017), Hajikarimi et al. (2018a) introduced the generalized fractional viscoelastic model by connecting simple fractional elements in series or parallel, as depicted in Fig. 3.17. As mentioned in Chapter 2, Constitutive Viscoelastic Functions, the total creep compliance can be calculated by the summation of creep compliance of all elements when the viscoelastic elements connect in series (see Fig. 3.17):

$$D(t) = \sum_{i=1}^{n} A_i t^{\alpha_i} \tag{3.65}$$

in which A_i and α_i are constant parameters for each simple fractional element and t is time. It is well-known that for the use of a viscoelastic model to characterize a specific material like bitumen, asphalt mastic, or asphalt mixture, it is required to be able to determine both creep

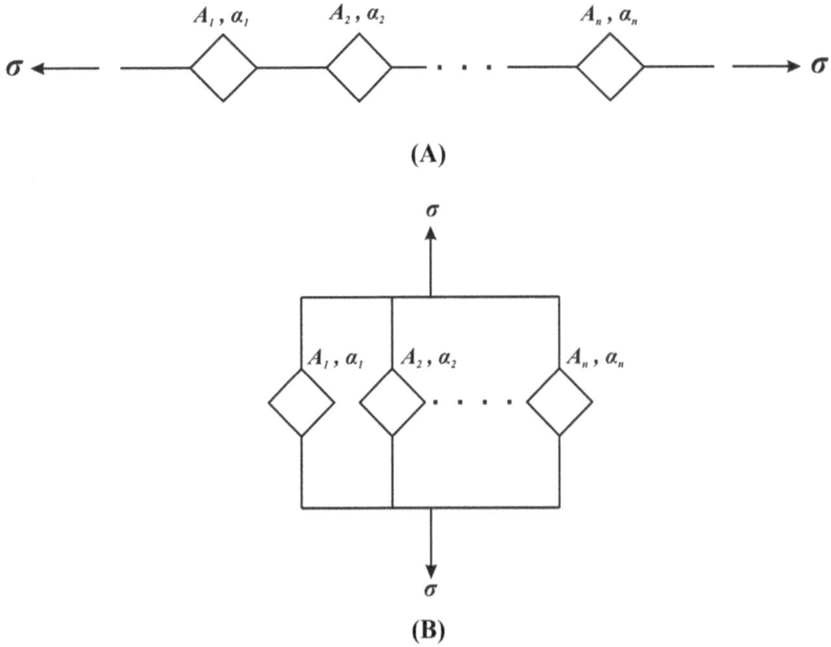

Figure 3.17 Generalized fractional viscoelastic model in (A) series and (B) parallel.

compliance and relaxation modulus. The main relation between these constitutive viscoelastic functions [$D(t)$ and $E(t)$] can be represented as follows, which will be more discussed in Chapter 6, Interconversion of Constitutive Viscoelastic Functions:

$$\int_0^t E(t-\xi)D(\xi)d\xi = t \tag{3.66}$$

By using the convolution theorem, Eq. (3.66) can be rewritten in the Laplace domain as follows:

$$\overline{E}(s)\overline{D}(s) = \frac{1}{s^2} \tag{3.67}$$

The Laplace transformation of the creep compliance can be derived as follows:

$$\overline{D}(s) = \sum_{i=1}^{n} A_i s^{-1-\alpha_i}\Gamma(1+\alpha_i) \tag{3.68}$$

and then by using Eq. (3.67), the relaxation modulus can be written in the Laplace domain:

$$\overline{E}(s) = \sum_{i=1}^{n} \frac{1}{A_i s^{1-\alpha_1} \Gamma(1+\alpha_i)} \tag{3.69}$$

where Γ is Gamma function, as defined in Eq. (3.53). By using the Laplace inverse transform, the relaxation modulus can be derived in the time domain as follows:

$$E(t) = \sum_{i=1}^{n} \frac{t^{-\alpha_i}}{A_i \Gamma(1-\alpha_i)\Gamma(1+\alpha_i)} \tag{3.70}$$

Similar to the procedure used for the simple fractional element based on Eqs. (2.24) and (2.25), the storage modulus and loss modulus of the generalized fractional viscoelastic model can be determined as:

$$|E'(\omega)| = \sum_{i=1}^{n} \frac{\omega^{\alpha_i}}{A_i \Gamma(1+\alpha_i)} \cos\left(\frac{\pi\alpha_i}{2}\right) \tag{3.71}$$

$$|E''(\omega)| = \sum_{i=1}^{n} \frac{\omega^{\alpha_i}}{A_i \Gamma(1+\alpha_i)} \sin\left(\frac{\pi\alpha_i}{2}\right) \tag{3.72}$$

The damping ratio can be determined based on Eq. (3.60) by dividing the loss modulus on the storage modulus. Furthermore, in order to determine the dissipated energy ratio based on Eq. (3.63), it is required to determine both stored and dissipated energy. In the following, the stored and dissipated energy relations for a generalized fractional viscoelastic model are represented:

$$W_s(t) = \sum_{i=1}^{n} A_i \tau_0^2 \left[t^{\alpha_i} - \frac{1}{2}(2t)^{\alpha_i} \right] \tag{3.73}$$

$$W_d(t) = \sum_{i=1}^{n} A_i \tau_0^2 \left[\frac{1}{2}(2t)^{\alpha_i} \right] \tag{3.74}$$

in which τ_0 is creep stress, A_i and α_i are constant parameters of the generalized fractional viscoelastic model, and t is time.

3.3.4 General fractional viscoelastic modeling

The fractional differential time equation (FDTE) can be represented as Eq. (3.75) for various combinations of springs and dashpots, as depicted in Fig. 3.18.

$$\sum_{j=1}^{4} b_j \frac{d^{\beta_j}}{dt^{\beta_j}} \tau = \sum_{j=1}^{4} c_j \frac{d^{\beta_j}}{dt^{\beta_j}} \gamma \tag{3.75}$$

The coefficients of Eq. (3.75) are presented in Table 3.2 including β_j, b_j, and c_j ($j = 1$, 2, 3, and 4). Since the general model consists of two springs and three dashpots, consistent with de Benedetto's nomenclature, it is called the 2S3P element. Several well-known models which are used in the modeling of viscoelastic materials and originally are based on combination of springs and simple/fractional dashpots can be simply derived by applying some assumptions on the 2S3P element. In Table 3.2, a_T denotes the temperature shift factor which can be calculated from different types of equations (see Chapter 5: Time–Temperature Superposition). For example, the Williams–Landel–Ferry (WLF) (Williams et al., 1955) equation is presented in the following:

$$\log(a_T) = \frac{-c_1\left(T - T_{ref}\right)}{c_2 + \left(T - T_{ref}\right)} \tag{3.76}$$

where T_{ref} is the reference temperature, T is the viscoelastic material temperature, c_1 and c_2 are constant specific parameters determined based on the results of experiments at different test temperatures.

A harmonic loading function [similar to Eq. (2.17) for strain] for stress can be represented as follows:

$$\tau(t) = \tau_0 \cdot e^{i\omega t} \tag{3.77}$$

in which τ_0 is the peak of stress, i denotes a complex number ($i^2 = -1$), ω is angular frequency ($\omega = 2\pi f$), and t is time. The strain response, $\gamma(t)$ due to this harmonic loading can be described as:

$$\gamma(t) = \gamma_0 \cdot e^{(i\omega t - i\delta)} \tag{3.78}$$

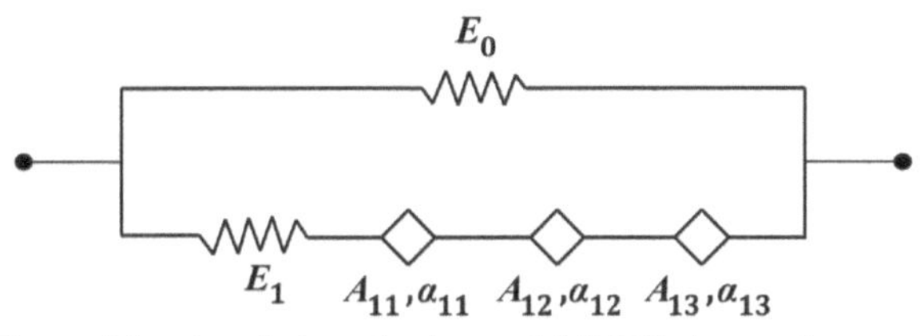

Figure 3.18 General fractional viscoelastic model (2S3P element).

Table 3.2 Exponents, stress, and strain coefficients of the 2S3P element.

Exponents	Stress coefficients	Strain coefficients
$\beta_1 = a_{11} + a_{12}$	$b_1 = \frac{E_1}{A_{13}.a_T{}^{a_{13}}}$	$c_1 = \frac{E_0.E_1}{A_{13}.a_T{}^{a_{13}}}$
$\beta_2 = a_{11} + a_{13}$	$b_2 = \frac{E_1}{A_{12}.a_T{}^{a_{12}}}$	$c_2 = \frac{E_0.E_1}{A_{13}.a_T{}^{a_{13}}}$
$\beta_3 = a_{12} + a_{13}$	$b_3 = \frac{E_1}{A_{11}.a_T{}^{a_{11}}}$	$c_3 = \frac{E_0.E_1}{A_{11}.a_T{}^{a_{11}}}$
$\beta_4 = a_{11} + a_{12} + a_{13}$	$b_4 = 1$	$c_4 = E_0 + E_1$

where γ_0 is the peak of strain and δ is phase angle (time lag between application of stress and strain response). The fractional time derivatives of both applied stress and strain response are given as:

$$\frac{d^\alpha}{dt^\alpha}\tau(t) = \tau_0 \cdot e^{i\omega t} \cdot e^{i\alpha\pi/2} \cdot \omega^\alpha \tag{3.79}$$

$$\frac{d^\alpha}{dt^\alpha}\gamma(t) = \gamma_0 \cdot e^{i\omega t} \cdot e^{i\alpha\pi/2} \cdot \omega^\alpha \cdot e^{-i\delta} \tag{3.80}$$

By substituting the fractional differential operators $(d^{\beta j}/dt^{\beta j})$ mentioned in Eq. (3.75) into Eqs. (3.79) and (3.80), the FDTE of the general model can be solved as:

$$\tau_0 \cdot \sum_{j=1}^{4}\left(b_j \cdot e^{i\beta_j\pi/2} \cdot \omega^{\beta_j}\right) = \gamma_0 \cdot \sum_{j=1}^{4}\left(c_j \cdot e^{i\beta_j\pi/2} \cdot \omega^{\beta_j}\right) \cdot e^{-i\delta} \tag{3.81}$$

Considering $(\tau_0/\gamma_0) = E^* \cdot e^{-i\delta}$, it is possible to rewrite Eq. (3.81) as follows:

$$E^* = \frac{\sum_{j=1}^{4}\left(c_j \cdot e^{i\beta_j\pi/2} \cdot \omega^{\beta_j}\right)}{\sum_{j=1}^{4}\left(b_j \cdot e^{i\beta_j\pi/2} \cdot \omega^{\beta_j}\right)} \tag{3.82}$$

The exponential function of $e^{i\beta_j\pi/2}$ can be represented as:

$$e^{i\beta_j\pi/2} = \cos\left(\frac{\beta_j\pi}{2}\right) + i \cdot \sin\left(\frac{\beta_j\pi}{2}\right) \tag{3.83}$$

which yields:

$$E^* = \frac{a_1 + i \cdot a_2}{a_3 + i \cdot a_4} \tag{3.84}$$

in which,

$$a_1 = \sum_{j=1}^{4}\left(c_j \cdot \cos\left(\frac{\beta_j \pi}{2}\right) \cdot \omega^{\beta_j}\right) \tag{3.85}$$

$$a_2 = \sum_{j=1}^{4}\left(c_j \cdot \sin\left(\frac{\beta_j \pi}{2}\right) \cdot \omega^{\beta_j}\right) \tag{3.86}$$

$$a_3 = \sum_{j=1}^{4}\left(b_j \cdot \cos\left(\frac{\beta_j \pi}{2}\right) \cdot \omega^{\beta_j}\right) \tag{3.87}$$

$$a_4 = \sum_{j=1}^{4}\left(b_j \cdot \sin\left(\frac{\beta_j \pi}{2}\right) \cdot \omega^{\beta_j}\right) \tag{3.88}$$

The complex modulus (E^*) can be written in the following format:

$$E^* = E' + i \cdot E'' \tag{3.89}$$

where E' denotes the storage modulus as:

$$E' = \frac{a_1 \cdot a_3 + a_2 \cdot a_4}{(a_3)^2 + (a_4)^2} \tag{3.90}$$

and E'' is the loss modulus as:

$$E'' = \frac{a_2 \cdot a_3 - a_1 \cdot a_4}{(a_3)^2 + (a_4)^2} \tag{3.91}$$

3.3.5 Fractional viscoelastic models for bituminous materials

There are several fractional viscoelastic models used in various material modeling fields. Among these fractional models, in this section four analogical models are introduced that are commonly used for bituminous materials, including Huet (1965), Sayegh (1967), 1S2P1D (Olard, 2003), and 2S2P1D (Olard & Di Benedetto, 2003). All these models can be derived by using the 2S3P element introduced in Section 3.3.4 as a general model. In the following, the abovementioned are briefly represented and compared with each other.

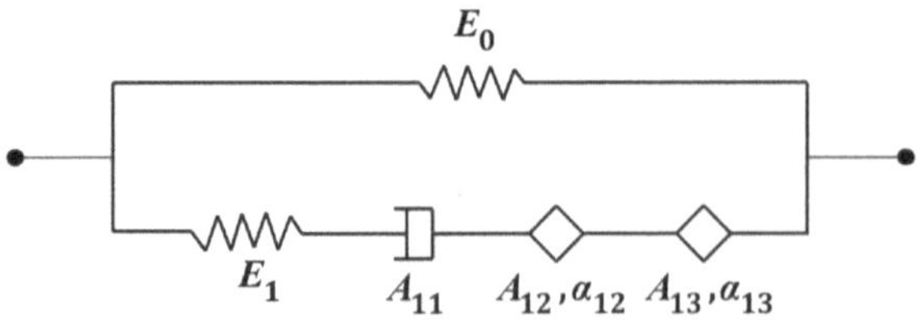

Figure 3.19 The 2S2P1D model.

Table 3.3 Coefficients of the 2S2P1D model.

Exponents	Stress coefficients	Strain coefficients
$\beta_1 = 1 + a_{12}$	$b_1 = \frac{E_1}{A_{13}.a_T{}^{a_{13}}}$	$c_1 = \frac{E_0.E_1}{A_{13}.a_T{}^{a_{13}}}$
$\beta_2 = 1 + a_{13}$	$b_2 = \frac{E_1}{A_{12}.a_T{}^{a_{12}}}$	$c_2 = \frac{E_0.E_1}{A_{12}.a_T{}^{a_{12}}}$
$\beta_3 = a_{12} + a_{13}$	$b_3 = \frac{E_1}{A_{11}.a_T{}^{a_{11}}}$	$c_3 = \frac{E_0.E_1}{A_{11}.a_T{}^{a_{11}}}$
$\beta_4 = 1 + a_{12} + a_{13}$	$b_4 = 1$	$c_4 = E_0 + E_1$

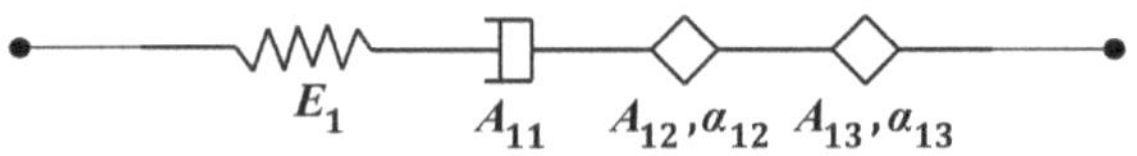

Figure 3.20 The 1S2P1D model.

3.3.5.1 2S2P1D model

If the first fractional dashpot in the 2S3P element is considered to be a linear dashpot ($\alpha_{11} = 1$), the 2S2P1D model will be derived as can be seen in Fig. 3.19. Table 3.3 shows coefficients of this model, which was successfully used by Olard and Di Benedetto (2003) for modeling bituminous materials.

3.3.5.2 1S2P1D model

By eliminating the single spring of the 2S2P1D model, another type of fractional viscoelastic model will be derived called 1S2P1D, as shown in Fig. 3.20. Table 3.4 shows the coefficients of the 1S2P1D model.

3.3.5.3 Huet–Sayegh model

By eliminating one of the fractional dashpots of the 2S3P element, the Huet–Sayegh model will be derived shown in Fig. 3.21. Table 3.5 shows the coefficients of the Huet–Sayegh model.

Table 3.4 Coefficients of the 1S2P1D model.

Exponents	Stress coefficients	Strain coefficients
$\beta_1 = 1 + a_{12}$	$b_1 = \frac{E_1}{A_{13}.a_T{}^{a_{13}}}$	$c_1 = 0$
$\beta_2 = 1 + a_{13}$	$b_2 = \frac{E_1}{A_{12}.a_T{}^{a_{12}}}$	$c_2 = 0$
$\beta_3 = a_{12} + a_{13}$	$b_3 = \frac{E_1}{A_{11}.a_T{}^{a_{11}}}$	$c_3 = 0$
$\beta_4 = 1 + a_{12} + a_{13}$	$b_4 = 1$	$c_4 = 0$

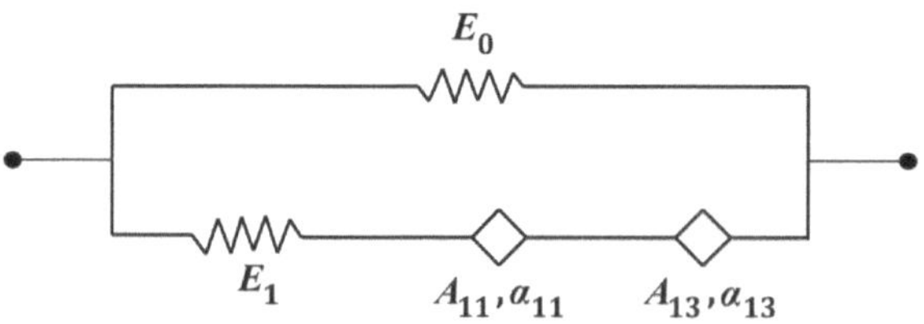

Figure 3.21 The Huet–Sayegh model.

Table 3.5 Coefficients of the Huet–Sayegh model.

Exponents	Stress coefficients	Strain coefficients
$\beta_1 = a_{11}$	$b_1 = \frac{E_1}{A_{13}.a_T{}^{a_{13}}}$	$c_1 = \frac{E_0.E_1}{A_{13}.a_T{}^{a_{13}}}$
$\beta_3 = a_{13}$	$b_3 = \frac{E_1}{A_{11}.a_T{}^{a_{11}}}$	$c_3 = \frac{E_0.E_1}{A_{11}.a_T{}^{a_{11}}}$
$\beta_3 = a_{11} + a_{13}$	$b_4 = 1$	$c_4 = E_0 + E_1$

3.3.5.4 Huet model

By eliminating the single spring of the Huet–Sayegh model, the Huet model will be given, which is depicted in Fig. 3.22. Table 3.6 shows the coefficients of the Huet model.

3.3.5.5 Comparing 2S2P1D, 1S2P1D, Huet–Sayegh, and Huet models

In order to compare different types of fractional viscoelastic models, including 2S2P1D, 1S2P1D, Huet–Sayegh, and Huet models, the Cole–Cole diagram can be used. Generally for viscoelastic materials like bitumen, asphalt mastic, asphalt mortar, and asphalt mixture, the Cole–Cole diagram is used to investigate the storage and loss modulus when subjected to harmonic loading. Fig. 3.23 depicts a Cole–Cole plot of an original bitumen along with fabricated asphalt mastic considering two different volume fractions of filler: 18% and 35%. As can be seen in

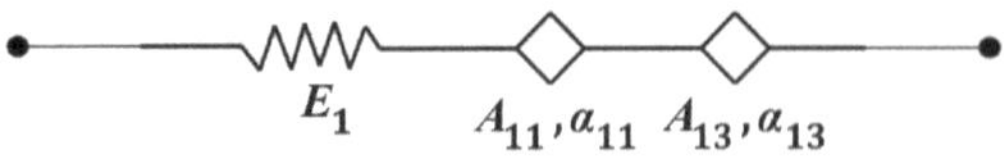

Figure 3.22 The Huet model.

Table 3.6 Coefficients of the Huet model.

Exponents	Stress coefficients	Strain coefficients
$\beta_1 = a_{11}$	$b_1 = \frac{E_1}{A_{13}.a_T{}^{a_{13}}}$	$c_1 = 0$
$\beta_3 = a_{13}$	$b_3 = \frac{E_1}{A_{11}.a_T{}^{a_{11}}}$	$c_3 = 0$
$\beta_3 = a_{11} + a_{13}$	$b_4 = 1$	$c_3 = E_1$

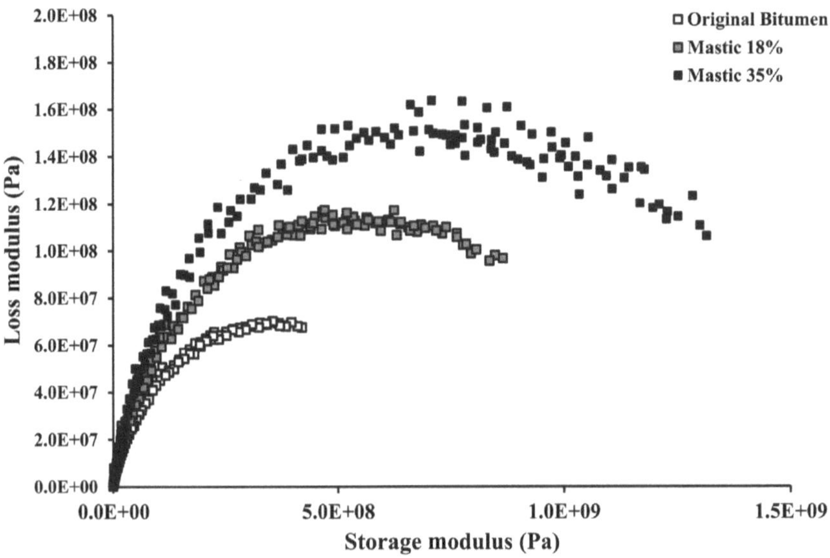

Figure 3.23 A sample Cole–Cole diagram for an original bitumen and asphalt mastic fabricated with two different volume fractions of mineral filler. *From https://doi.org/10.1061/(ASCE)MT.1943-5533.0002548.*

Fig. 3.23, loss modulus (E'') is plotted versus storage modulus (E') in a Cole–Cole diagram. Fig. 3.24 shows schematic Cole–Cole plots for 2S2P1D, 1S2P1D, Huet–Sayegh, and Huet models. Having a single dashpot causes a shift equal to E_0 at the left-side of the Cole–Cole plot of both the 1S2P1D and Huet–Sayegh models while both 1S2P1D and Huet models start at zero point. Additionally, to better illustrate these model together, Fig. 3.25 shows creep and relaxation behavior of these

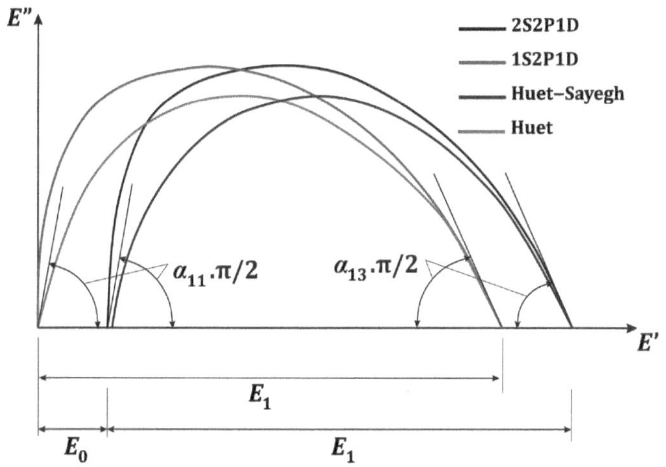

Figure 3.24 The Cole–Cole diagram of 2S2P1D, 1S2P1D, Huet–Sayegh, and Huet models.

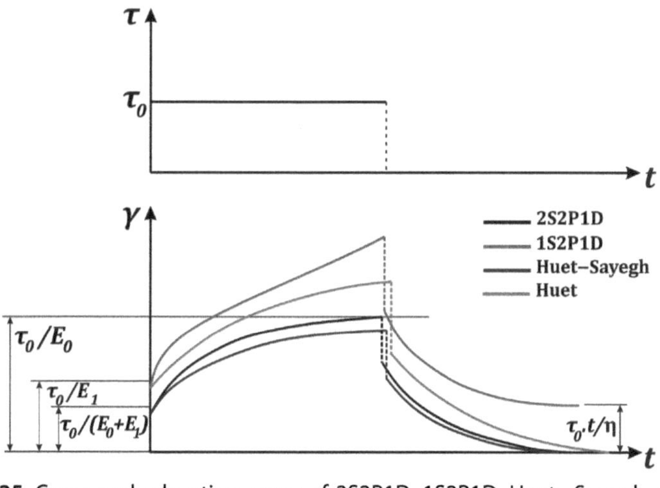

Figure 3.25 Creep and relaxation curve of 2S2P1D, 1S2P1D, Huet–Sayegh, and Huet models.

models schematically during application of a constant stress (τ_0) and its removal. Among all these models, it can be seen that the 2S2P1D model demonstrates viscous strain that will not recover with time and is capable of simulating permanent deformation. Also, it is evident that there is a limitation for the strain of both 2S2P1D and Huet–Sayegh models, which is specified by the spring capacity as shown in this Fig. 3.25.

References

Aflaki, S., & Hajikarimi, P. (2012). Implementing viscoelastic rheological methods to evaluate low temperature performance of modified asphalt binders. *Construction and Building Materials, 36*, 110−118. Available from https://doi.org/10.1016/j.conbuildmat.2012.04.076.

Bagley, R. L., & Torvik, P. (1983). A theoretical basis for the application of fractional calculus to viscoelasticity. *Journal of Rheology, 27*(3), 201−210.

Brinson, H. F., & Brinson, L. C. (2008). *Polymer engineering science and viscoelasticity*. Springer.

Burgers, J. M. (1935). *First and second report on viscosity and plasticity*. the Netherlands: Academy of Sciences at Amsterdam.

Caputo, M., & Mainardi, F. (1971). A new dissipation model based on memory mechanism. *Pure and Applied Geophysics, 91*(1), 134−147. Available from https://doi.org/10.1007/BF00879562.

Gemant, A. (1936). A method of analyzing experimental results obtained from elasto-viscous bodies. *Physics*, 7(8), 311−317.

Graziani, A., Mignini, C., Bocci, E., & Bocci, M. (2020). Complex modulus testing and rheological modeling of cold-recycled mixtures. *Journal of Testing and Evaluation, 48*(1).

Gutierrez-Lemini, D. (2014). *Engineering viscoelasticity*. New York: Springer.

Hajikarimi, P., Rahi, M., & Moghadas Nejad, F. (2015). Comparing different rutting specification parameters using high temperature characteristics of rubber-modified asphalt binders. *Road Materials and Pavement Design, 16*(4), 751−766.

Hajikarimi, P., Fakhari Tehrani, F., Moghadas Nejad, F., Absi, J., Rahi, M., Khodaii, A., & Petit, C. (2018a). Generalized fractional viscoelastic modeling of low temperature characteristics of asphalt binders modified with polyphosphoric acid and distillate aromatic extracts oil. *Journal of Materials in Civil Engineering, 30*(7), Article 04018147. Available from https://doi.org/10.1061/(ASCE)MT.1943-5533.0002353.

Hajikarimi, P., Moghadas Nejad, F., & Mohammadi Aghdam, M. (2018b). Implementing general power law to interconvert linear viscoelastic functions of modified asphalt binders. *Journal of Transportation Engineering, Part B: Pavements, 144*(2), Article 04018010. Available from https://doi.org/10.1061/JPEODX.0000038.

Huet, C. (1965). Etude par une méthode d'impédance du comportement viscoélastique des matériaux hydrocarbonés.

Kelvin, W. (1875). On the theory of viscoelastic fluids. *Mathematical Physics Papers, 3*, 27−84.

Mangiafico, S., Sauzéat, C., & Di Benedetto, H. (2019). 2S2P1D model calibration error from user panel for one bitumen and one bituminous mixture. *Advances in Materials Science and Engineering*, 2019.

Maxwell, J. C. (1867). IV. On the dynamical theory of gases. *Philosophical Transactions of the Royal Society of London, 157*, 49−88. Available from https://doi.org/10.1098/rstl.1867.0004.

Nutting, P. (1921). A new general law of deformation. *Journal of the Franklin Institute, 191*(5), 679−685.

Olard, F. (2003). *Comportement thermomécanique des enrobés bitumineux à basses températures. Relations entre les propriétés du liant et de l'enrobé*. Lyon: INSA de Lyon.

Olard, F., & Di Benedetto, H. (2003). General "2S2P1D" model and relation between the linear viscoelastic behaviours of bituminous binders and mixes. *Road Materials and Pavement Design, 4*(2), 185−224.

Ottosen, N. S., & Ristinmaa, M. (2005). *The mechanics of constitutive modeling*. Elsevier.

Park, S. W., & Kim, Y. R. (1999). Interconversion between relaxation modulus and creep compliance for viscoelastic solids. *Journal of Materials in Civil Engineering, 11*(1), 76−82.

Sayegh, G. (1967). Viscoelastic properties of bituminous mixtures. *The International Conference on Structural Design of Asphalt Pavements.*
Stimilli, A., Virgili, A., Canestrari, F., & Bahia, H. U. (2017). Estimation of low-temperature performance of recycled asphalt mixtures through relaxation modulus analysis. *Cold Regions Science and Technology*, *133*, 36–45. Available from https://doi.org/10.1016/j.coldregions.2016.10.013.
Tschoegl, N. W. (1989). *The phenomenological theory of linear viscoelastic behavior*. New York: Springer.
Voigt, W. (1892). Ueber innere Reibung fester Körper, insbesondere der Metalle. *Annalen Der Physik*, *283*(12), 671–693. Available from https://doi.org/10.1002/andp.18922831210.
Williams, M. L., Landel, R. F., & Ferry, J. D. (1955). The temperature dependence of relaxation mechanisms in amorphous polymers and other glass-forming liquids. *Journal of the American Chemical Society*, 77(14), 3701–3707. Available from https://doi.org/10.1021/ja01619a008.
Woldekidan, M., Huurman, M., & Pronk, A. (2012). A modified HS model: Numerical applications in modeling the response of bituminous materials. *Finite Elements in Analysis and Design*, *53*, 37–47.
Xu, Q., & Solaimanian, M. (2009). Modelling linear viscoelastic properties of asphalt concrete by the Huet–Sayegh model. *International Journal of Pavement Engineering*, *10*(6), 401–422.
Xu, Y., Shan, L., & Tian, S. (2019). Fractional derivative viscoelastic response model for asphalt binders. *Journal of Materials in Civil Engineering*, *31*(6), Article 04019089.

CHAPTER 4

Correspondence principle of viscoelasticity

4.1 Introduction

There is a main concern to find closed form solutions (the relation between stresses, strains, and displacements, as described in Chapter 1, Introduction to Viscoelasticity) for problems in which materials behave as linear or nonlinear viscoelastic substances, which means that their behavior depends on time, temperature, and loading rate. There are several classic and prevalent problems with closed form solutions in which the main assumption is the elastic behavior of materials, like the Boussinesq problem (Podio-Guidugli & Favata, 2014), for determining the elastic state in a linearly elastic isotropic half-space, subject to a concentrated load applied in a point of its boundary plane and perpendicular to it, or a simple three-point bending beam considering the Euler–Bernoulli or the Timoshenko theorem assumptions. Since the main assumptions of an elastic solution do not exist for a viscoelastic problem, it is not possible to directly use classical and widespread elastic solutions for the problems concerning viscoelastic materials. Using the correspondence principle (CP), which is also called Alfrey's correspondence principle (Brinson & Brinson, 2008), is a well-known appropriate method to implement elastic solutions for finding viscoelastic ones by using the Laplace transformation properties. In this chapter, first the basic theoretical framework of the CP is presented, and then some simple useful examples are given, including three-point bending beam, axially loaded bar, an Euler–Bernoulli with loading conditions other than bending, and displacement domain near a crack tip in fracture mechanics. Although it is claimed that any viscoelastic relation can be derived from its corresponding elastic relation based on the CP, using the CP is not straightforward for all problems in the viscoelastic domain. It is necessary to mention that to follow the content presented in this chapter needs a preliminary background in Laplace transformation (Schiff, 2013) and methods of resolving differential equations (Boyce et al., 2017).

Applications of Viscoelasticity
DOI: https://doi.org/10.1016/B978-0-12-821210-3.00007-3

4.2 Theoretical background on the correspondence principle

As mentioned in Sections 3.2.5 and 3.3.4, the general format of the constitutive integer-order or fractional-order equation of a viscoelastic material can be represented as (Ottosen & Ristinmaa, 2005):

$$\sum_{i=0}^{n} a_i \frac{d^i \sigma}{dt^i} = \sum_{j=1}^{m} b_j \frac{d^j \varepsilon}{dt^j} \tag{4.1}$$

in which a_i and b_j are constant parameters and generally m and n can be different. Considering Eq. (4.1), it is evident that it is a linear differential equation and hence the superposition principle can be used. It means that if a stress history of $\sigma = \sigma_1(t)$ leads to a strain of $\varepsilon_1(t)$, then a stress history of $\sigma = k\sigma_1(t)$ will result in a strain of $k\varepsilon_1(t)$. Additionally, if a stress history of $\sigma = \sigma_1(t)$ leads to a strain of $\varepsilon_1(t)$ and another stress history of $\sigma = \sigma_2(t)$ leads to a strain of $\varepsilon_2(t)$, then the total stress history $[\sigma(t) = \sigma_1(t) + \sigma_2(t)]$ will result in a strain of $\varepsilon(t) = \varepsilon_1(t) + \varepsilon_2(t)$. This is the simple description of the superposition principle that can be used for linear viscoelastic materials and problems containing such substances. To simplify Eq. (4.1), it is possible to rewrite it in a condensed format as follows:

$$P\sigma = Q\varepsilon \tag{4.2}$$

where P and Q are operators that can be defined as:

$$P = \sum_{i=0}^{n} a_i \frac{d^i}{dt^i} \tag{4.3}$$

$$Q = \sum_{j=1}^{m} b_j \frac{d^j}{dt^j} \tag{4.4}$$

The Laplace transform of an arbitrary function of $f(t)$, which is defined as follows (Boyce et al., 2017):

$$L\{f(t)\} = \bar{f}(s) = \int_0^{\infty} \exp(-st) f(t) dt \tag{4.5}$$

can be simply used to convert differential equations to algebraic equations. The Laplace transformation of a function's derivation can be represented based on Eq. (4.5) as:

$$L\{f'(t)\} = \int_0^{\infty} \frac{d[f(t)]}{dt} e^{-st} dt = s\bar{f}(s) - f(0) \tag{4.6}$$

By applying the Laplace transform of Eq. (4.5) on Eq. (4.1) and considering the Laplace transformation of the derivative of a function [Eq. (4.6)], the following relation can be derived:

$$\sum_{i=0}^{n} a_i s^i \overline{\sigma}(s) = \sum_{j=1}^{m} b_j s^j \overline{\varepsilon}(s) \tag{4.7}$$

in which s is the Laplace variable and the relation of stress and strain in the transformed Laplace domain is in the form of an algebraic equation. This equation can be condensed similar to Eq. (4.2) as:

$$\overline{P}(s)\overline{\sigma}(s) = \overline{Q}(s)\overline{\varepsilon}(s) \tag{4.8}$$

in which $\overline{P}(s)$ and $\overline{Q}(s)$ are defined as:

$$\overline{P}(s) = \sum_{i=0}^{n} a_i s^i \tag{4.9}$$

$$\overline{Q}(s) = \sum_{j=1}^{m} b_j s^j \tag{4.10}$$

Finally, Eq. (4.8) can be rewritten in the following format:

$$\overline{\sigma}(s) = \left(\frac{\sum_{i=0}^{n} a_i s^i}{\sum_{j=1}^{m} b_j s^j} \right) \overline{\varepsilon}(s) = \frac{\overline{Q}(s)}{\overline{P}(s)} \overline{\varepsilon}(s) \tag{4.11}$$

which is similar to Hook's law (Timoshenko & Young, 1962) in the elastic domain for axial loading that can be represented as:

$$\overline{\sigma}(s) = \overline{E}^*(s)\overline{\varepsilon}(s) \tag{4.12}$$

or

$$\overline{\varepsilon}(s) = \overline{D}^*(s)\overline{\sigma}(s) \tag{4.13}$$

Both Eqs. (4.12) and (4.13) are known as Alfrey's correspondence principle (CP) equations, in which stress and strain in the Laplace transform domain are connected via $\overline{E}*(s)$ and $D*(s)$. The concept of converting viscoelastic problems in the time domain into elastic problems in the transform domain was first introduced and discussed by Alfrey (1944). Later in 1950 and in 1955, the principle was generalized and discussed by

Read (1950) and Lee (1955), respectively. As can be seen in Eq. (4.12), $\overline{E}*(s)$ in this equation is similar to Young's modulus for a linear elastic material. By using the CP, the linear relation between stress and strain of a viscoelastic material in the integral form [see Eqs. (2.7) and (2.14)] in the time domain can be converted to a linear elastic form in the Laplace transform domain.

On the other hand, considering the hereditary approach for explaining the mechanical behavior of viscoelastic materials, it is possible to apply the Laplace transform on Eq. (2.14) as:

$$L\{\sigma(t)\} = L\left\{\int_0^t E(t-\xi)\frac{d\varepsilon(\xi)}{d\xi}d\xi\right\} \tag{4.14}$$

where $L\{.\}$ denotes Laplace transformation. The Laplace transform of a convolution integral can be determined as (Boyce et al., 2017):

$$L\left\{\int_0^t f(t-\eta)g(\eta)d\eta\right\} = L\left\{\int_0^t f(\eta)g(t-\eta)d\eta\right\} = L\{f(t)\}L\{g(t)\} \tag{4.15}$$

Therefore Eq. (4.14) can be rewritten as:

$$\begin{aligned} L\{\sigma(t)\} = L\{E(t)\}\,.\,L\left\{\frac{d\varepsilon(t)}{dt}\right\} &= L\{E(t)\}\,.\,[s\,L\{\varepsilon(t)\} - \varepsilon(0)] \\ &= s\,L\{E(t)\}L\{\varepsilon(t)\} \end{aligned} \tag{4.16}$$

where $\varepsilon(0) = 0$. Eq. (4.16) can be represented as follows:

$$\overline{\sigma}(s) = s\,\overline{E}(s)\overline{\varepsilon}(s) \tag{4.17}$$

Furthermore, by applying the Laplace transform on Eq. (2.7), and performing a similar approach, the following equation can be derived simply:

$$\overline{\varepsilon}(s) = s\,\overline{D}(s)\overline{\sigma}(s) \tag{4.18}$$

Therefore by comparing Eq. (4.17) with Eq. (4.12) and also Eq. (4.18) with Eq. (4.13), it can be concluded that:

$$\overline{E}^*(s) = s\,\overline{E}(s) \tag{4.19}$$

and

$$\overline{D}^*(s) = s\,\overline{D}(s) \tag{4.19}$$

It is necessary to mention that both the differential equation approach and hereditary approach are two methods for explaining the mechanical behavior of viscoelastic materials and there is no inherent difference between these approaches. By substituting the right side of Eq. (4.18) into Eq. (4.17), the following equation can be simply derived:

$$\overline{E}(s)\,\overline{D}(s) = \frac{1}{s^2} \tag{4.20}$$

which can be inversely transformed to the time domain by considering the convolution theorem as follows:

$$\int_0^t E(t)D(t-\xi)d\xi = t \tag{4.21}$$

The abovementioned equation apparently shows that the relaxation modulus is not equal to the inverse of creep compliance and generally the following equation must be considered between $E(t)$ and $D(t)$ in the time domain (Flügge, 2013).

$$E(t) \neq \frac{1}{D(t)} \tag{4.22}$$

However, these two constitutive functions have a direct relation in the Laplace transform domain, as can be seen in Eq. (4.23):

$$\overline{D}^*(s) = \frac{1}{\overline{E}^*(s)} \tag{4.23}$$

It is necessary to mention that both Eqs. (4.21) and (4.23) can be used for determining one of the main constitutive viscoelastic functions based on another one. This procedure, which is called interconversion between viscoelastic functions, will be presented and discussed in Chapter 6, Interconversion of Constitutive Viscoelastic Functions, considering different methods.

Having derived equations for relating stresses, strains, displacements, and constitutive functions (relaxation modulus and creep compliance) in the time domain to the Laplace transform domain and vice versa, it is possible to define a step-by-step procedure for applying the CP in order to develop a solution for a viscoelastic problem in the time domain based on an existing elastic solution. This procedure is described in the following (Brinson & Brinson, 2008):

Step 1: Find an analytical solution for a linear elastic boundary value problem with geometry, loading, and boundary conditions which are exactly identical to the linear viscoelastic boundary value problem that one desires to find the solution.

Step 2: Substitute all variables of the analytical elastic solution including stresses, strains, displacements, and applied loads with their Laplace transform.

Step 3: Substitute all elastic constants (like elastic modulus and Poisson's ratio) with s times the transform of the time-dependent moduli as:

$$
\begin{aligned}
E &\rightarrow \overline{E}^*(s) = s\overline{E}(s) \\
D &\rightarrow \overline{D}^*(s) = s\overline{D}(s) \\
\nu &\rightarrow \overline{\nu}^*(s) = s\overline{\nu}(s)
\end{aligned}
\tag{4.24}
$$

Step 4: Apply inverse Laplace transformation to find the desired viscoelastic solution in the time domain.

In order to show the ability of the CP and provide a better insight for using this principle to derive linear viscoelastic solutions, some useful examples presented in the next sections.

4.3 Examples of using the correspondence principle

In this section, three different examples are presented to show how the correspondence principle (CP) can be used to find a solution that is needed for a linear viscoelastic boundary value problem. All examples are based on a previously solved linear elasticity boundary value problem which can be simply found in the literature.

4.3.1 Viscoelastic three-point bending beam

To investigate the low-temperature mechanical behavior of original and modified bitumens as well as asphalt mastic, a well-known test setup of bending beam rheometer (BBR) (Bahia, Anderson, & Christensen, 1992) is used which is shown in Figure 2.2. As can be seen in this figure, this is a simple three-point bending test setup in which a constant load of ($P_0 = 0.98$ N) is applied at the mid-span of the beam with dimensions of $127 \times 12.7 \times 6.35$ mm. The main output of the BBR test is mid-span deflection during the loading time (240 s) which then should be used to calculate flexural creep stiffness and creep rate as two main criteria to evaluate low-temperature behavior of bituminous materials (ASTM D6648-08, 2001). Since bituminous materials are considered viscoelastic material, it is required to find the time-dependent relation of mid-span deflection. According to the elementary bending theory (Armenàkas, 2016) for an

Euler–Bernoulli beam, for an elastic material, the maximum deflection at the mid-span can be calculated as:

$$\delta = \frac{PL^3}{48EI} \tag{4.25}$$

where δ is deflection of beam at mid-span, P the applied constant load, L the span length, E the modulus of elasticity, I the moment of inertia of the section $= bh^3/12$, and b the width and h the depth of the beam, respectively. Having a viscoelastic beam causes a time-dependent behavior for mid-span deflection, increasing for a constant applied load (creep condition). By using the step-by-step procedure introduced in Section 4.2, it is possible to find the solution for a viscoelastic beam. At first, the deflection and applied load are substituted with their Laplace transforms. Also, the elastic modulus of E is substituted with $s\overline{E}(s)$. The following equation is derived in the Laplace transform domain:

$$\overline{\delta}(s) = \frac{\overline{P}(s)L^3}{48s\overline{E}(s)I} \tag{4.26}$$

It is necessary to mention that both L and I are dependent on geometry of the beam and no adjustment is required for these two parameters. As previously mentioned, the applied load in a BBR test is constant [$P(t) = P_0$] and its Laplace transform is:

$$\overline{P}(s) = P_0/s \tag{4.27}$$

which can be substituted in Eq. (4.26) as follows:

$$\overline{\delta}(s) = \frac{P_0L^3}{48I}\frac{1}{\left(s^2\overline{E}(s)\right)} \tag{4.28}$$

With regard to Eq. (4.20), it is possible to rewrite Eq. (4.28) as:

$$\overline{\delta}(s) = \frac{P_0L^3\overline{D}(s)}{48I} \tag{4.29}$$

which can be simply transformed to the time domain by applying the inverse Laplace transform as:

$$\delta(t) = \frac{P_0L^3D(t)}{48I} \tag{4.30}$$

where $D(t)$ is creep compliance that can be determined with mid-span deflections as the main output of a BBR test recorded each 0.5 s.

The inverse of creep compliance is called flexural creep stiffness [$S(t) = 1/D(t)$] and Eq. (4.30) can be reorganized as:

$$S(t) = \frac{P_0 L^3}{4bh^3 \delta(t)} \tag{4.31}$$

as it is introduced in ASTM D6648. There is a common mistake for calculating the relaxation modulus of bituminous materials at low-temperature where some researchers used $S(t)$ as a relaxation modulus, although it is not true because $E(t) \neq 1/D(t)$. Hajikarimi, Nejad, & Aghdam (2018) investigated such a difference to simulate the numerical behavior of a bending beam in comparison with experimental results to show that there is a significant difference between $S(t)$ and $E(t)$ even at low test temperatures. In order to calculate the relaxation modulus based on creep compliance, it is required to use an interconversion method which is introduced in Chapter 6, Interconversion of Constitutive Viscoelastic Functions.

4.3.2 Viscoelastic axially loaded bar

Fig. 4.1 shows elastic and viscoelastic axially loaded bars. The average or engineering stress and the average or engineering strain of the axially loaded bar shown in Fig. 4.1 can be calculated as (Armenàkas, 2016):

$$\sigma(t) = \frac{P(t)}{A} \tag{4.32}$$

$$\varepsilon(t) = \frac{\psi(t)}{L} \tag{4.33}$$

where $\sigma(t)$ is stress, $\varepsilon(t)$ is strain, $P(t)$ is the applied load, A is the original cross-sectional area, L is the original bar length, and E is the Young's modulus which is constant for the elastic bar. It is evident that the stress

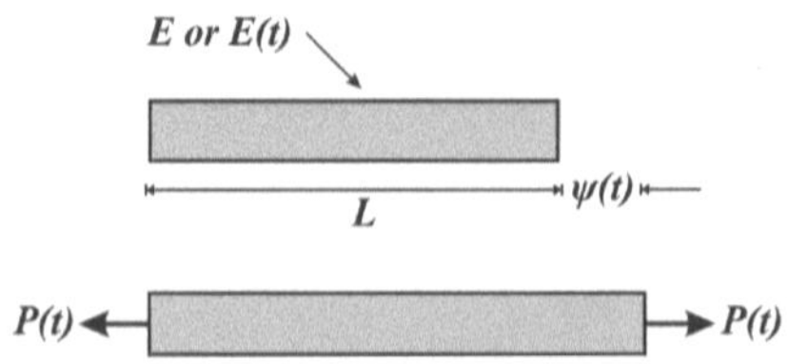

Figure 4.1 Axially loaded elastic/viscoelastic bar.

and strain can be related based on Hook's law as it is represented in Eq. (4.34):

$$\sigma(t) = E\varepsilon(t) \tag{4.34}$$

By substituting Eqs. (4.32) and (4.33) within Eq. (4.34), the axial deformation is found as:

$$\frac{P(t)}{A} = E\frac{\psi(t)}{L} \quad \Rightarrow \quad \psi(t) = \frac{P(t)L}{EA} \tag{4.35}$$

As can be seen in Eq. (4.35), the main reason for the variation of the deformation is the variation of applied load. One may focus on the variation of cross-sectional area with time. If such a variation is considered to determine the stress, it will result in true stress which is approximately identical with average stress for small deformations.

To find an analytical solution for an axially loaded viscoelastic bar, the step-by-step procedure of using the correspondence principle is again used. For this purpose, the deformation, $\psi(t)$, and the applied load, $P(t)$, are substituted with their Laplace transforms as well as Young's modulus as follows:

$$\overline{\psi}(s) = \frac{\overline{P}(s)L}{s\overline{E}(s)A} \xrightarrow{s\overline{E}(s) = \frac{1}{s\overline{D}(s)}} \overline{\psi}(s) = \frac{\overline{P}(s)L}{A} s\overline{D}(s) \tag{4.36}$$

In order to apply the inverse Laplace transform, it is possible to use the convolution theorem, and the time-dependent deformation is derived as:

$$\psi(t) = \frac{L}{A}\int_0^t D(t-\xi)\frac{dP(\xi)}{d\xi}d\xi \tag{4.37}$$

Having creep compliance and applied load function, one can choose to use Eq. (4.36) or Eq. (4.37) with regard to the complexity of the problem. For a simple case of loading in which the axial load is a constant step, such as $P(t) = P_0\ H(t)$ where $H(t)$ is Heaviside function, the Laplace transform is as follows:

$$\overline{P}(s) = \frac{P_0}{s} \tag{4.38}$$

and the deformation can be derived as:

$$\psi(t) = \frac{L}{A}\int_0^t D(t-\xi)\frac{d[P_0 H(\xi)]}{d\xi}d\xi = \frac{P_0 L}{A}\int_0^t D(t-\xi)\delta(\xi)d\xi \tag{4.39}$$

in which $\delta(\xi)$ is the Dirac delta function. Eq. (4.39) can be represented as:

$$\psi(t) = \frac{P_0 L}{A} D(t) \tag{4.40}$$

For another case of applied load in which $P(t) = P_0 t$, its Laplace transform would be:

$$\overline{P}(s) = \frac{P_0}{s^2} \tag{4.41}$$

Substituting Eq. (4.41) into Eq. (4.39) shows that it is necessary to select an appropriate viscoelastic mechanical model to derive the closed form solution for a viscoelastic bar. For example, considering a Kelvin–Voigt model for describing mechanical behavior of a viscoelastic material and substituting Eq. (3.12) as the creep compliance function into Eq. (4.39), the following equation can be derived:

$$\psi(t) = \frac{P_0 L}{EA}\left(t - \frac{\eta}{E}\left(1 - \exp\left[-\frac{E}{\eta}t\right]\right)\right) \tag{4.42}$$

4.3.3 Viscoelastic beams for conditions other than bending

In this section, some instances are represented for a simply supported viscoelastic beam with different loading conditions.

4.3.3.1 Uniformly distributed loading

Fig. 4.2 shows a viscoelastic beam with an uniformly distributed load. In order to use the CP for finding the stresses, deformations, and strain of this viscoelastic beam, an elastic beam with the same geometry and loading is considered.

The maximum bending stress is captured at the outer surface of the beam's mid-span as (Armenàkas, 2016):

$$\sigma_{\max} = \frac{M_{\max} y}{I} = \frac{3}{2}\frac{w(t)L^2}{h^3} = \frac{3}{2}\frac{w_0 L^2}{h^3} \tag{4.43}$$

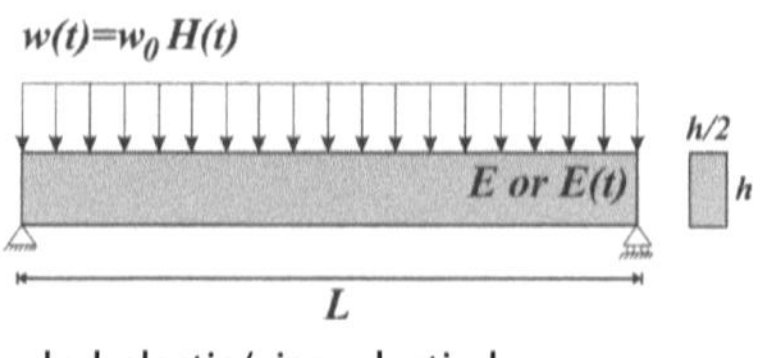

Figure 4.2 Uniformly loaded elastic/viscoelastic beam.

in which M is the bending moment, I is the moment of inertia (second moment of area), and y is the vertical distance from the neutral axis to the location in the cross section where the stress is to be measured. To conduct the CP, it is required to replace the stress and the uniform load with their Laplace transforms. However, there is no need to replace geometry properties including L and h, as mentioned in Section 4.2. Eq. (4.43) can be represented in the transformed Laplace domain as:

$$\overline{\sigma}_{\max}(s) = \frac{3}{2}\frac{\overline{w}(s)L^2}{h^3} \xrightarrow{w(t)=w_o H(t)} \overline{\sigma}_{\max}(s) = \frac{3}{2}\frac{w_0 L^2}{sh^3} \tag{4.44}$$

and then the inversion of this solution will give the desired solution in the time domain as:

$$\sigma_{\max}(t) = \frac{3}{2}\frac{w(t)L^2}{h^3} \tag{4.45}$$

which is identical with the elastic solution. It can be stated that if there is no elastic constant existing in the elastic solution, then it can be expected that the corresponding viscoelastic solution will be exactly the same. However, the appearance of elastic constants in the mechanical expressions changes the story, as stated earlier for determining the strain of an axially loaded bar (Section 4.3.2).

Additionally, the maximum deflection of the elastic beam is:

$$\psi_{\max} = \frac{5}{348}\frac{w_0 L^4}{EI} \tag{4.46}$$

Again, by using the CP and replacing all variables with their Laplace transforms as well as writing elastic constants of E as $s\bar{E}(s)$, Eq. (4.46) can be represented in the Laplace transform domain as:

$$\overline{\psi}_{\max}(s) = \frac{5}{348}\frac{\overline{w}(s)L^4}{s\overline{E}(s)I} \xrightarrow{s\overline{E}(s)=\frac{1}{s\overline{D}(s)}} \overline{\psi}_{\max}(s) = \frac{5}{348}\frac{\overline{w}(s)L^4}{I}s\overline{D}(s) \tag{4.47}$$

Considering the Laplace transform for $w(t)$, the inversion of Eq. (4.47) into the time domain would be:

$$\psi_{\max}(t) = \frac{5}{348}\frac{w_0 L^4}{I}D(t) \tag{4.48}$$

where $D(t)$ is a creep compliance function and is directly dependent on the mechanical behavior considered for describing viscoelastic material. For example, if the Burgers model is assumed for a viscoelastic material,

Eq. (3.15) can be substituted in Eq. (4.48) as *D(t)*. Then, the maximum deflection would be:

$$\psi_{\max}(t) = \frac{5}{348}\frac{w_0 L^4}{I}\left[\frac{1}{E_M} + \frac{t}{\eta_M} + \frac{1}{E_K}\left(1 - \exp\left(-\frac{E_K}{\eta_K}t\right)\right)\right] \tag{4.49}$$

4.3.3.2 Arbitrary distributed loading (function of distance and time)

Fig. 4.3 depicts a beam loaded with a distributed load which is a function of both distance and time. The general form of such loading can be expressed as $P(x,t)$ while in this section a specific case is considered in which $P(x,t) = P(x)F(t)$. Actually, the dependency of loading function to distance and time is separated and they are independent of each other.

For implementing the CP, the elastic solution for an elastic beam with the same geometry and loading can be written as:

$$\psi(x,t) = \frac{P(x,t)}{EI}g(x) = \frac{P(x)F(t)}{EI} \tag{4.50}$$

in which $g(x)$ contains spatial distribution of the deflection solution for $P(x)$. To determine $g(x)$, the deflection for an elastic beam with a uniformly varying loading of $P(x,t) = p_0 x$ can be considered as follows:

$$\psi(x) = \frac{1}{EI}\frac{p_0 x}{360L}\left(7L^4 - 10L^2x^2 + 3x^4\right) \tag{4.51}$$

By comparing Eqs. (4.50) and (4.51), one can simply realize that $g(x)$ for such a loading condition would be:

$$g(x) = \left(7L^4 - 10L^2x^2 + 3x^4\right) \tag{4.52}$$

Now, replacing all variables in the Eq. (4.51) with their Laplace transforms based on the CP leads to:

$$\overline{\psi}(x,s) = \frac{P(x)g(x)}{I}\frac{\overline{F}(s)}{s\overline{E}(s)} \xrightarrow{s\overline{E}(s) = \frac{1}{s\overline{D}(s)}} \overline{\psi}(x,s) = \frac{P(x)g(x)}{I}\left[\overline{F}(s).s\overline{D}(s)\right] \tag{4.53}$$

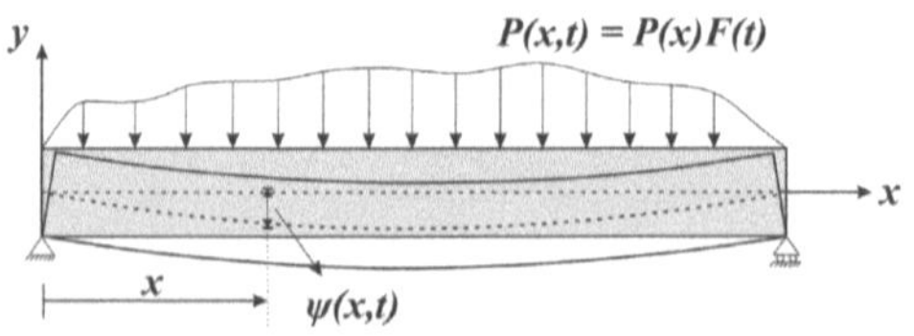

Figure 4.3 Spatially varying distributed load on a viscoelastic beam.

and then by using the definition of convolution integral, the desired solution in the time domain will be:

$$\psi(x,t) = \frac{P(x)g(x)}{I}\int_{0^-}^{t} D(t-\xi)\frac{dF(\xi)}{d\xi}d\xi \tag{4.54}$$

For the loading shown in Fig. 4.3, the integral must include the jump discontinuity at the origin by knowing the initial value of $F(t)$ as $F_0H(t)$, which results in the following equation for $\psi(t)$:

$$\psi(x,t) = \frac{P(x)g(x)}{I}\left[F_0D(t) + \int_{0^+}^{t} D(t-\xi)\frac{dF(\xi)}{d\xi}d\xi\right] \tag{4.55}$$

4.3.4 Linear viscoelastic fracture mechanics

In this section, by using the Alfrey's correspondence principle, the analytical solution for two-dimensional linear viscoelastic fracture mechanics is derived for determining a displacement field around the crack tip for both Mode I (plane crack opening) and Mode II (in-plane shear) of fracture, shown in Fig. 4.4 (Broek, 2012). Here, the main assumption is that the Poisson's ratio of the viscoelastic material is not dependent on the time and remains constant during loading and unloading. It is an acceptable assumption for low- and moderate-performance temperature for bituminous materials as an example of linear viscoelastic materials (Al-Qadi et al., 2010; Di et al., 2007).

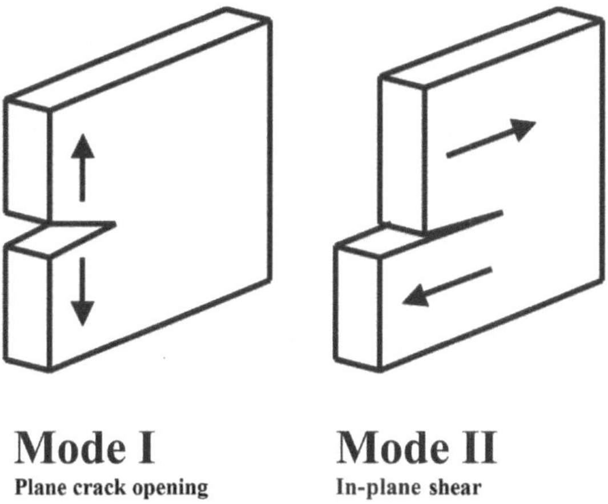

Figure 4.4 Mode I (plane crack opening) and Mode II (in-plane shear) of fracture.

According to linear elastic fracture mechanics, the displacement field around the crack tip for Mode I of opening is as follows (Saouma, 2000):

$$u_x(r,\theta) = \frac{K_I}{2G}\sqrt{\frac{r}{2\pi}}\cos\frac{\theta}{2}(\kappa - \cos\theta) \tag{4.56}$$

$$u_y(r,\theta) = \frac{K_I}{2G}\sqrt{\frac{r}{2\pi}}\sin\frac{\theta}{2}(\kappa - \cos\theta) \tag{4.57}$$

in which $G = E/2(1+\nu)$ is shear modulus, K_I is the stress intensity factor (SIF) of the first mode of fracture, and κ is the Kolosov constant which can be calculated based on the Poisson's ratio (ν) for both plane stress and plane strain conditions, respectively, as:

$$\kappa = \frac{3-\nu}{1+\nu} \tag{4.58}$$

$$\kappa = 3 - 4\nu \tag{4.59}$$

By using the four-step procedure of Alfrey's correspondence application described in Section 4.2, it is possible to rewrite Eqs. (4.56) and (4.57) in the Laplace transform domain by replacing all variables with their Laplace transforms as:

$$\overline{u}_x(s) = \frac{\overline{K}_I(s)}{2\overline{G}^*(s)}\sqrt{\frac{r}{2\pi}}\cos\frac{\theta}{2}(\overline{\kappa}^*(s) - \cos\theta) \tag{4.60}$$

$$\overline{u}_y(s) = \frac{\overline{K}_I(s)}{2\overline{G}^*(s)}\sqrt{\frac{r}{2\pi}}\sin\frac{\theta}{2}(\overline{\kappa}^*(s) - \cos\theta) \tag{4.61}$$

The applied load is considered as:

$$P(t) = P_0 f(t) \tag{4.62}$$

where P_0 is constant. Considering the applied load, the SIF can be simply derived as:

$$K_I(t) = K_I f(t) \rightarrow \overline{K}_I(s) = K_I \overline{f}(s) \tag{4.63}$$

where K_I is the SIF of an elastic material subjected to constant load of P_0. To proceed to the solution, it is required to consider a viscoelastic mechanical model to specify the relaxation modulus function. For example, the three parameter standard solid viscoelastic model is considered here, which is the simplest form of the generalized Maxwell model (Fig. 3.12) previously introduced in Chapter 3, Mechanical Models of Viscoelasticity. This model is depicted in Fig. 4.5. With regard to Eq. (3.25), the relaxation modulus of the three parameter standard solid viscoelastic model is:

$$E(t) = E_\infty + E_1 \exp\left(-\frac{E_1}{\eta_1}t\right) \tag{4.64}$$

The Young's modulus, E, and the shear modulus, μ, are related for an elastic material as:

$$G = \frac{E}{2(1+\nu)} \tag{4.65}$$

In order to use the correspondence principle, the elastic constants can be replaced with their Laplace transforms by remembering that the Poisson's ratio is assumed to be constant. Therefore Eq. (4.65) can be written in the Laplace transform domain as:

$$\overline{G}^*(s) = \frac{\overline{E}^*(s)}{2(1+\nu)} \tag{4.66}$$

Now, it is required to substitute the Laplace transform of Eq. (4.64) into Eq. (4.66) which can be determined as:

$$\overline{E}(s) = L\left\{E(t)\right\} = L\{E_\infty\} + L\left\{\left[E_1 \exp\left(-\frac{E_1}{\eta_1}t\right)\right]\right\} = \frac{E_\infty}{s} + \frac{E_1}{s + \frac{E_1}{\eta_1}} \tag{4.67}$$

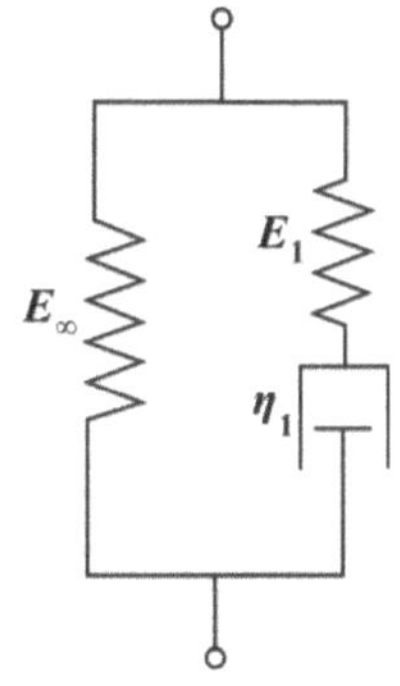

Figure 4.5 Three parameter standard viscoelastic model.

and $\bar{E}^*(s)$ can be simply determined as:

$$\overline{E}^*(s) = s\overline{E}(s) = E_\infty + \frac{sE_1}{s + \frac{E_1}{\eta_1}} \tag{4.68}$$

The last variable needed to be determined is the Laplace transform of the Kolosov constant. With regard to assuming the Poisson's ratio as a constant parameter, the Laplace transform of the Kolosov constant can be derived as:

$$\overline{\kappa}^*(s) = s\overline{\kappa}(s) = s\frac{\kappa}{s} = \kappa \tag{4.69}$$

Now, by substituting Eq. (4.63), Eq. (4.66), and Eq. (4.69) into Eq. (4.60) and Eq. (4.61), the displacement field near the crack tip for Mode I of two-dimensional fracture can be written in the Laplace transform domain as:

$$\overline{u}_x(s) = \frac{K_I\overline{f}(s)\,(1+\nu)}{\left[E_\infty + \frac{sE_1}{s+\frac{E_1}{\eta_1}}\right]}\sqrt{\frac{r}{2\pi}}\cos\frac{\theta}{2}(\kappa - \cos\theta) \tag{4.70}$$

$$\overline{u}_y(s) = \frac{K_I\overline{f}(s)\,(1+\nu)}{\left[E_\infty + \frac{sE_1}{s+\frac{E_1}{\eta_1}}\right]}\sqrt{\frac{r}{2\pi}}\sin\frac{\theta}{2}(\kappa - \cos\theta) \tag{4.71}$$

and then, by considering the simplest creep loading condition of $f(t) = 1$, the inverse Laplace transform can be applied on Eqs. (4.70) and (4.71) to achieve the following equations in the time domain:

$$u_x(r,\theta,t) = K_I(1+\nu)\left(\frac{1}{E_\infty} - \frac{E_1}{E_\infty(E_\infty + E_1)}\exp\left(-t/\tau'\right)\right)$$
$$\sqrt{\frac{r}{2\pi}}\cos\frac{\theta}{2}(\kappa - \cos\theta) \tag{4.72}$$

$$u_y(r,\theta,t) = K_I\,(1+\nu)\left(\frac{1}{E_\infty} - \frac{E_1}{E_\infty(E_\infty + E_1)}\exp\left(-t/\tau'\right)\right)$$
$$\sqrt{\frac{r}{2\pi}}\sin\frac{\theta}{2}(\kappa - \cos\theta) \tag{4.73}$$

where $\tau' = \left(\frac{1}{E_\infty} + \frac{1}{E_1}\right)\eta_1$.

Both Eqs. (4.72) and (4.73) can be represented in the following condensed form based on the stress intensity factor:

$$u_x(r,\theta,t) = (1+\nu)\varphi(t)K_I(t)\sqrt{\frac{r}{2\pi}}\cos\frac{\theta}{2}(\kappa - \cos\theta) \tag{4.74}$$

$$u_y(r,\theta,t) = (1+\nu)\varphi(t)K_I(t)\sqrt{\frac{r}{2\pi}}\sin\frac{\theta}{2}(\kappa - \cos\theta) \tag{4.75}$$

in which,

$$\varphi(t) = \frac{1}{E_\infty} - \frac{E_1}{E_\infty(E_\infty + E_1)}\exp\left(-t/\tau'\right) \tag{4.76}$$

Using a similar procedure, the following equations can be found for the displacement field near a crack tip in Mode II of fracture for a two-dimensional medium:

$$u_x(r,\theta,t) = (1+\nu)\varphi(t)K_{II}(t)\sqrt{\frac{r}{2\pi}}\sin\frac{\theta}{2}(\kappa + \cos\theta + 2) \tag{4.77}$$

$$u_y(r,\theta,t) = -(1+\nu)\varphi(t)K_{II}(t)\sqrt{\frac{r}{2\pi}}\cos\frac{\theta}{2}(\kappa + \cos\theta - 2) \tag{4.78}$$

in which $K_{II}(t)$ is the stress intensity factor (SIF) for Mode II of fracture.

As can be seen in the abovementioned formulation, the main drawback of this procedure is that for a more complex viscoelastic mechanical model like the generalized Maxwell model with more components or the generalized Kelvin–Voigt model, inversing the Laplace transform equations into the time domain will be complex and in some cases no closed form solution can be derived.

As mentioned in Section 3.3, it is possible to use fractional-order viscoelastic models to describe the mechanical behavior of viscoelastic materials. In order to show how such models can be used to simplify the derivation of closed form viscoelastic solutions, a simple fractional viscoelastic model introduced in Section 3.3.2 is used for deriving the displacement field near a crack tip for a two-dimensional linear viscoelastic fracture mechanics problem. For a simple fractional element, the relaxation modulus can be represented as:

$$E(t) = At^{-\alpha} \tag{4.79}$$

in which A and α are real numbers so that $A>0$ and $0<\alpha<1$. By applying the Laplace transform on Eq. (4.79), it would be:

$$\overline{E}(s) = As^{\alpha-1}\Gamma(1-\alpha) \tag{4.80}$$

where Γ denotes the Gamma function which is defined as:

$$\Gamma(\alpha) = \int_0^{\infty} e^{-t} t^{\alpha-1} dt, \quad \mathrm{Re}(\alpha)>0 \tag{4.81}$$

It is evident that $\bar{E}^*(s)$ and $\bar{E}$ can be related via the following equation:

$$\overline{E}^*(s) = s\overline{E}(s) = As^{\alpha}\Gamma(1-\alpha) \tag{4.82}$$

and then by substituting Eq. (4.82) into Eq. (4.66) and using Eq. (4.60) and Eq. (4.61), the following equations can be written:

$$\overline{u}_x(s) = \frac{K_I\overline{f}(s)(1+\nu)}{As^{\alpha}\Gamma(1-\alpha)}\sqrt{\frac{r}{2\pi}}\cos\frac{\theta}{2}(\kappa-\cos\theta) \tag{4.83}$$

$$\overline{u}_y(s) = \frac{K_I\overline{f}(s)(1+\nu)}{As^{\alpha}\Gamma(1-\alpha)}\sqrt{\frac{r}{2\pi}}\sin\frac{\theta}{2}(\kappa-\cos\theta) \tag{4.84}$$

By implementing the inversion of the Laplace transform, it is possible to determine $u_x(t)$ and $u_y(t)$ for Mode I of fracture in the time domain as follows:

$$u_x(r,\theta,t) = (1+\nu)\chi(t)K_I(t)\sqrt{\frac{r}{2\pi}}\cos\frac{\theta}{2}(\kappa-\cos\theta) \tag{4.85}$$

$$u_y(r,\theta,t) = (1+\nu)\chi(t)K_I(t)\sqrt{\frac{r}{2\pi}}\sin\frac{\theta}{2}(\kappa-\cos\theta) \tag{4.86}$$

in which for creep loading of $f(t)=1$, the $\chi(t)$ would be:

$$\chi(t) = \frac{t^{\alpha}}{A\Gamma(\alpha-1)\Gamma(\alpha+1)} \tag{4.87}$$

and correspondingly it can be simply derived for Mode II of fracture:

$$u_x(r,\theta,t) = (1+\nu)\chi(t)K_{II}(t)\sqrt{\frac{r}{2\pi}}\sin\frac{\theta}{2}(\kappa+\cos\theta+2) \tag{4.88}$$

$$u_y(r,\theta,t) = (1+\nu)\chi(t)K_{II}(t)\sqrt{\frac{r}{2\pi}}\cos\frac{\theta}{2}(\kappa+\cos\theta-2) \tag{4.89}$$

References

Alfrey, T. (1944). Non-homogeneous stresses in visco-elastic media. *Quarterly of Applied Mathematics, 2*(2), 113–119.

Al-Qadi, I. L., Wang, H., & Tutumluer, E. (2010). Dynamic analysis of thin asphalt pavements by using cross-anisotropic stress-dependent properties for granular layer. *Transportation Research Record, 2154*(1), 156–163.

Armenàkas, A. E. (2016). *Advanced mechanics of materials and applied elasticity*. CRC Press.

Bahia, H. U., Anderson, D. A., & Christensen, D. W. (1992). The bending beam rheometer; a simple device for measuring low-temperature rheology of asphalt binders (with discussion). *Journal of the Association of Asphalt Paving Technologists, 61*, 17–153.

Boyce, W. E., DiPrima, R. C., & Meade, D. B. (2017). *Elementary differential equations*. John Wiley & Sons.

Brinson, H. F., & Brinson, L. C. (2008). *Polymer engineering science and viscoelasticity: An introduction*. Springer.

Broek, D. (2012). *Elementary engineering fracture mechanics*. Springer Science & Business Media.

D6648-08, A. (2001). *Standard test method for determining the flexural creep stiffness of asphalt binder using the bending beam rheometer (BBR)*.

Di, B. H., Brice, D., & Cédric, S. (2007). Three-dimensional linear behavior of bituminous materials: Experiments and modeling. *International Journal of Geomechanics*, 7(2), 149–157. Available from https://doi.org/10.1061/(ASCE)1532-3641(2007)7:2(149).

Flügge, W. (2013). *Viscoelasticity*. Springer Science & Business Media.

Hajikarimi, P., Nejad, F. M., & Aghdam, M. M. (2018). Implementing general power law to interconvert linear viscoelastic functions of modified asphalt binders. *Journal of Transportation Engineering Part B: Pavements, 144*(2), 682–687. Available from https://doi.org/10.1061/JPEODX.0000038.

Lee, E. H. (1955). Stress analysis in visco-elastic bodies. *Quarterly of Applied Mathematics, 13* (2), 183–190.

Ottosen, N. S., & Ristinmaa, M. (2005). *The mechanics of constitutive modeling*. Elsevier.

Podio-Guidugli, P., & Favata, A. (2014). In P. Podio-Guidugli, & A. Favata (Eds.), *The Boussinesq Problem BT - Elasticity for geotechnicians: A modern exposition of Kelvin, Boussinesq, Flamant, Cerruti, Melan, and Mindlin Problems* (pp. 79–114). Springer International Publishing. Available from https://doi.org/10.1007/978-3-319-01258-2_5.

Read, W. T. (1950). Stress analysis for compressible viscoelastic materials. *Journal of Applied Physics, 21*(7), 671–674. Available from https://doi.org/10.1063/1.1699729.

Saouma, V. E. (2000). *Fracture mechanics. Lecture notes*. Boulder: Dept. of Civil Environmental and Architectural Engineering, University of Colorado.

Schiff, J. L. (2013). *The Laplace transform: Theory and applications*. Springer Science & Business Media.

Timoshenko, S., & Young, D. H. (1962). *Elements of strength of materials*. Princeton: Van Nostrand.

CHAPTER 5

Time–temperature superposition

5.1 Introduction

The elastic materials are the most widely used materials for engineering applications. It is assumed that no inherent change occurs in their mechanical properties over time (Pelleg, 2012). Therefore there is no concern for determining properties of such materials regarding to time and temperature. The elastic materials can be recognized by having elasticity modulus (Young's modulus), E, and Poisson's ratio, ν which are not functions of any variable in the elastic domain (Sadd, 2009). While it is necessary to perform experiments at a specific temperature to determine these two parameters for an elastic material based on the standards, it is not common to report the test temperature for both design and construction applications. However, the story is different for viscoelastic materials as their mechanical properties directly depend on time, temperature, and loading rate (Tschoegl, 2012) and accordingly change over the life span of a material designed for a specific application. It means that it is necessary to consider this issue for determining constitutive functions of viscoelastic materials, including the creep compliance, $D(t)$, and the relaxation modulus, $E(t)$ (see Chapter 2: Constitutive Viscoelastic Functions). The notations that we usually used for these two functions neglect the effect of temperature and loading rate as it is assumed that they are presented for a specific temperature or loading rate. The thorough notation should be $D(t, T, \gamma)$ or $E(t, T, \gamma)$, where t is time (or frequency), T is temperature, and γ is loading rate.

There is a major limitation in determining viscoelastic properties over a wide range of time (or frequency) or temperature with regard to experimental equipment capability restrictions and the impossibility of some experiments at certain temperatures. On the other hand, improving the equipment capabilities to widen measurable time or temperature range significantly increases the cost of experiments. For example, the dynamic shear rheometer (DSR), which is a conventional device to measure viscoelastic properties of polymers and bituminous materials, can be used to perform different modes of experiments including frequency sweep,

Applications of Viscoelasticity
DOI: https://doi.org/10.1016/B978-0-12-821210-3.00006-1

temperature sweep, time sweep, and strain sweep at a frequency between 0.1 and 100 Hz. It is not possible to use this device to apply lower or higher frequencies on the specimens due to the instrumentation system of this device. Additionally, performing the DSR test is not possible for test temperatures higher than the melting point of polymers or around 90°C for bituminous materials because the specimen cannot withstand the test setup and immediately flows. Accordingly, accelerated test methods are required to evaluate the mechanical and rheological characteristics of viscoelastic materials at time (or frequency) and test temperatures in which it is not possible to measure them directly. Assuming the simple thermorheological behavior allows engineers to use the time−temperature superposition principle (TTSP), which is introduced and thoroughly explained in this chapter. In this regard, the concept of the master curve is introduced and some mathematical procedures are represented to construct the master curve at a reference temperature for a viscoelastic material.

5.2 Effect of temperature on viscoelastic properties

Considering a simple frequency sweep test performed on an original bitumen specimen (a disk with 8 mm diameter and 2 mm thickness) by implementing a dynamic shear rheometer at seven different test temperatures including −22°C, −16°C, −10°C, 0°C, 10°C, 16°C, and 22°C, it is possible to observe the effect of frequency and temperature on viscoelastic characteristics of the original bitumen. Fig. 5.1 shows the complex shear modulus (G^*) and phase angle (δ) as two main test outputs which are constitutive viscoelastic functions. As can be seen in Fig. 5.1A, increasing the test temperature decreases G^* and vice versa. Additionally, increasing the test frequency leads to increasing the complex shear modulus which apparently shows altering viscoelastic properties relating to time and temperature. For the phase angle, increasing the test temperature increases this parameter which is amplified by decreasing the test frequency. Fig. 5.1 presents the results of test temperatures both below and above zero, and it can be clearly seen that viscoelastic behavior in terms of dependency on time and temperature exists up to −22°C.

Fig. 5.2 shows flexural creep stiffness, $S(t)$, for an original bitumen specimen (a bending beam with 127 mm length, 12.7 mm width, and 6.35 thickness) at different test temperatures. This parameter was determined by implementing a bending beam rheometer and measuring midspan deflection of the bending beam every 0.5 s under a constant load of

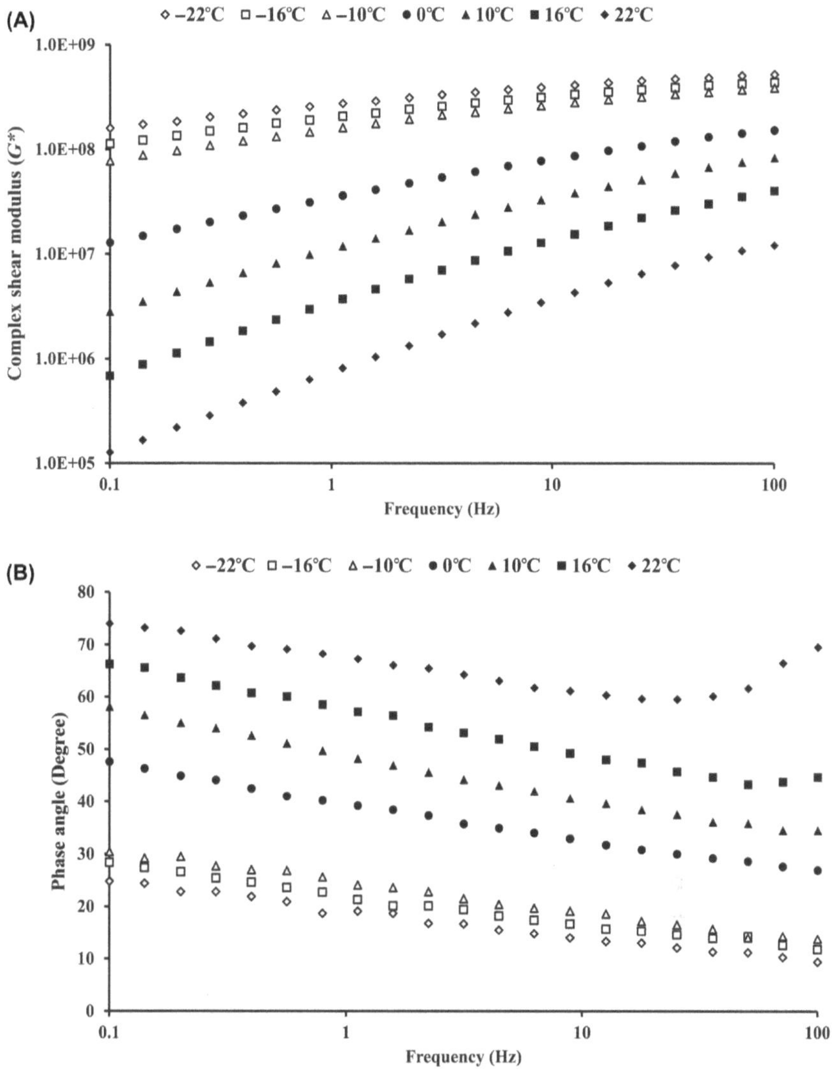

Figure 5.1 Results of frequency sweep test on an original bitumen at different frequencies and test temperatures, (A) complex shear modulus (G^*), and (B) phase angle (δ).

0.98 N (creep loading condition). The duration of loading is 240 s according to ASTM D6648. Fig. 5.2 illustrates that increasing the test temperature decreases the $S(t)$ which means that bitumen will soften at higher test temperatures.

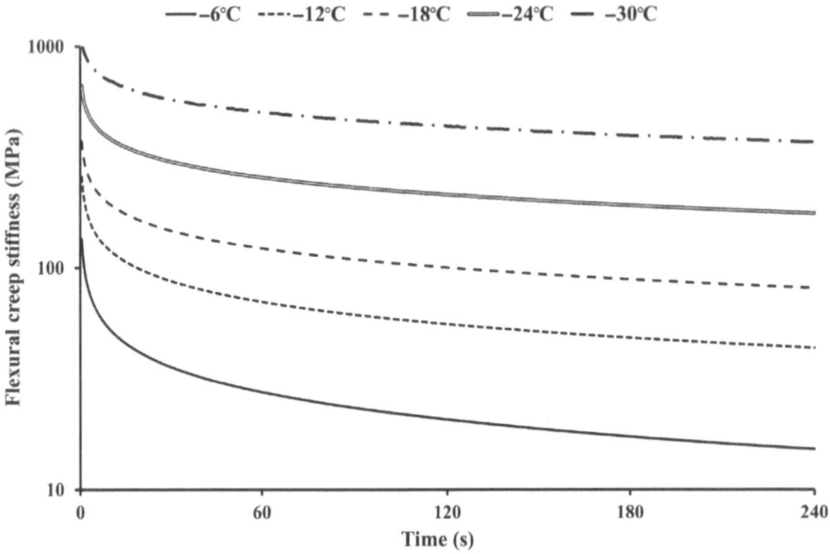

Figure 5.2 Results of flexural creep stiffness of an original bitumen at various test temperatures.

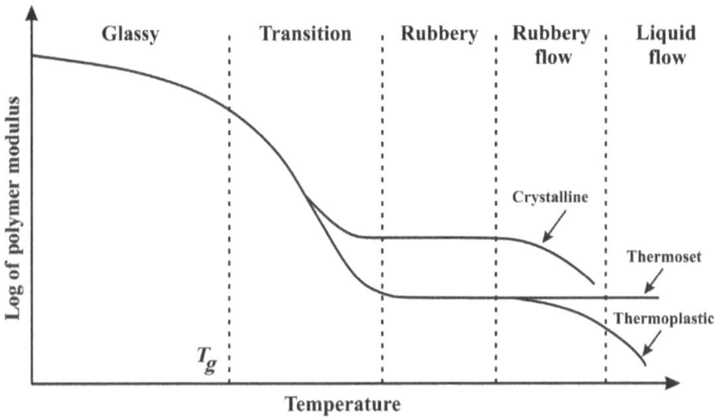

Figure 5.3 Schematic viscoelastic behavior of polymers at various temperatures.

A general schematic viscoelastic behavior of a polymer can be represented as shown in Fig. 5.3 which illustrates that the modulus of polymers, such as viscoelastic materials, remarkably differs with temperature. The difference between crystalline, thermoset, and thermoplastic can be seen in this figure; these are defined in relation to a polymer's temperature-dependent behavior. The glass transition temperature, T_g, is also defined as a temperature at which the amorphous regions experience a transition from rigid state to more flexible state making the temperature at the border of the solid state to a rubbery state.

5.3 Time–temperature superposition principle to develop master curves

Fig. 5.4 illustrates using the time–temperature superposition principle (TTSP) to develop a master curve at a reference temperature of 0°C for the original bitumen's complex shear modulus (G^*) results depicted in Fig. 5.1A. As can be seen in this figure, the curves related to test temperatures lower than 0°C were shifted to the right, while those related to test temperatures higher than 0°C were shifted to the left to form a continuous curve called a "master curve." Forming a master curve by shifting the curves related to various test temperatures to the right and left demonstrates that the viscoelastic material can be categorized as a thermorheologically simple material. In other words, if a viscoelastic material is thermorheologically simple, the distance by which a given value like complex shear modulus or any other viscoelastic functions at a temperature of T must be translated horizontally to fall onto the curve at the reference temperature, T_{ref}, is the same for all values of frequency or time.

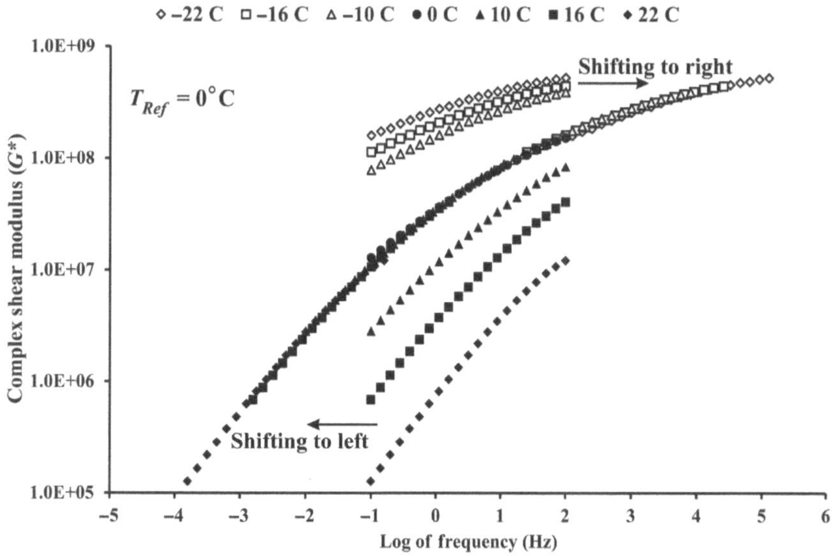

Figure 5.4 Master curve construction for complex shear modulus (G^*) of an original bitumen at a reference temperature of 0°C.

Some major points can be found by looking at the master curve constructed in Fig. 5.4: (1) the master curve just represents the rheological behavior of the original bitumen at a specific and desired test temperature called "reference temperature" (T_{ref}). This temperature must be selected from among all test temperatures for which the considered viscoelastic function is measured or determined. For example, all seven test temperatures shown in Fig. 5.4 can be a candidate for the reference temperature but $T = -20°C$ cannot be set as T_{ref}.

(2) It can be seen that while the frequency sweep experiments were performed in a frequency range between 0.1 and 100 Hz for all test temperatures, the master curve presented complex shear modulus data in a frequency range between 0.0001 and 80,000 Hz which apparently shows widening the frequency range to overcome experimental apparatus limitations and the test setup as previously described in Section 5.1. The curves related to test temperatures lower than 0°C constructed the right part of the master curve, corresponding to higher frequencies, and the curves related to $T > 0°C$ formed the left part of the master curve, matching with lower frequencies. In other words, since it is not possible to achieve a higher frequency of 80,000 Hz by using a conventional dynamic shear rheometer in the laboratory at the test temperature of 0°C, it is simply possible to use results of other test temperatures to determine the desired viscoelastic function at very low/high frequencies or times without any further experiment.

(3) The 0°C master curve can be shifted to construct a new master curve at a desired temperature between −22°C and +22°C. Considering Fig. 5.4, if the 0°C master curve shifted to the left, a master curve can be formed for temperatures lower than 0°C, which alters the frequency range. On the other hand, for constructing master curves for temperatures higher than 0°C, it is possible to shift the 0°C master curve to the right. Altering the frequency range for the master curves at different temperatures shows the ability of TTSP to trade off temperature for frequency (or time). In other words, TTSP can be implemented to perform rheological or mechanical experiments at a limited frequency range with short duration at multiple temperatures to determine a viscoelastic function like complex shear modulus as a function of frequency (or time) spanning many decades in frequency or time.

Another example is presented in Fig. 5.5 of the application of TTSP on relaxation modulus data versus time for an original bitumen. All the abovementioned descriptions can be seen in this figure which shows

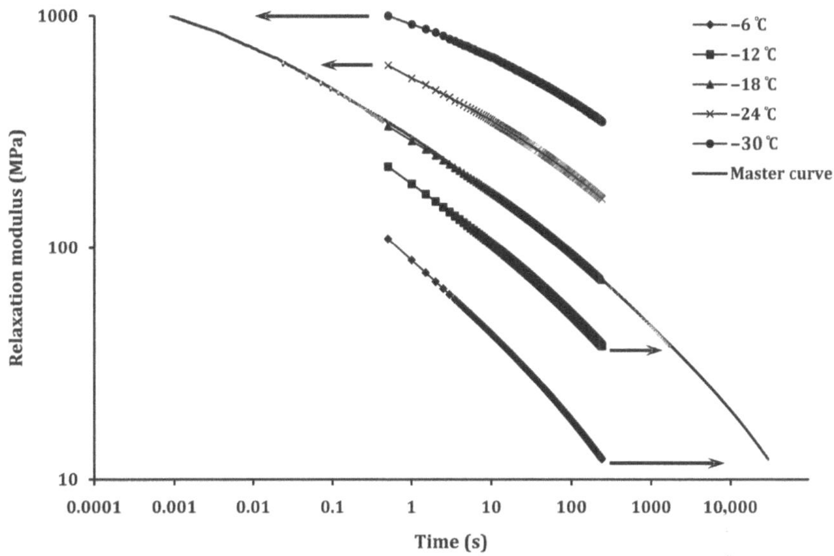

Figure 5.5 Master curve construction for relaxation modulus of an original bitumen at a reference temperature of −18°C.

construction of a master curve for a reference temperature of −18°C. It can be seen that while the bending beam rheometer (BBR) test was performed at each test temperature (−6°C, −12°C, −18°C, −24°C, and −30°C) for a short time of 240 s, the resulting master curve covers a time of 0.001–10,000 s.

5.4 Shift functions

As mentioned in Section 5.3, by using the TTSP it will be possible to construct a master curve at any temperature within the temperature range at which the experiment to measure the desired viscoelastic parameter or function is performed. The procedure of shifting is simple for test temperatures (like −22°C, −16°C, −10°C, 10°C, 16°C, and 22°C in Fig. 5.4) as there is a curve that one can see when the existing master curve relies on this curve. The main question is about constructing a master curve for other desired temperatures for which there is no experimental data, for example, if one aims to form a master curve for a temperature of 6°C based on the 0°C master curve of Fig. 5.4 or a master curve for the temperature of −20°C based on the −18°C master curve depicted in Fig. 5.5. How is it possible to determine the required amount of shifting

Table 5.1 Shift factors of an original bitumen to form a master curve at a reference temperature of 0°C.

Temperature (°C)	Log (a_T)	a_T
−22	3.00	1000.00
−16	2.50	316.23
−10	1.80	63.10
0	0	1.00
10	−1.00	0.1000
16	−1.80	0.0158
22	−2.70	0.0019

to the right or to the left to achieve the master curve? To answer these questions, two different shift functions are introduced in this section.

5.4.1 Williams–Landel–Ferry equation

In order to determine the shift factor for each temperature, Williams et al. (1955) introduced the following equation:

$$\log(a_T) = \frac{-c_1 \times \left(T - T_{Ref}\right)}{c_2 + \left(T - T_{Ref}\right)} \tag{5.1}$$

where a_T is the shift factor at temperature T, T_{ref} is the reference temperature for which a master curve was formed, and c_1 and c_2 are constant coefficients determined by fitting Eq. (5.1) on shift factors for test temperatures. For example, the shift factors of all test temperatures used to form the master curve shown in Fig. 5.4 are presented in Table 5.1. Fitting Eq. (5.1) on this data, c_1 and c_2 are determined as 27.45 and 209.3. Fig. 5.6 depicts shift factors as well as the Williams–Landel–Ferry (WLF) shift function fitted on this data.

Having a shift function like WLF, it is possible to determine the desired viscoelastic function at any temperature within the test temperature range. To show this procedure, T_a is assumed to be an arbitrary temperature. To better illustrate this procedure, a quantitative example is considered for the 0° master curve shown in Fig. 5.4 considering $T_a = 6°$ C. Log(a_T) can be simply determined by using the following equation:

$$\log(a_T) = \frac{-c_1 \times \left(T_a - T_{Ref}\right)}{c_2 + \left(T_a - T_{Ref}\right)} = \frac{-27.45 \times (6-0)}{209.3 + (6-0)} = -0.765 \tag{5.2}$$

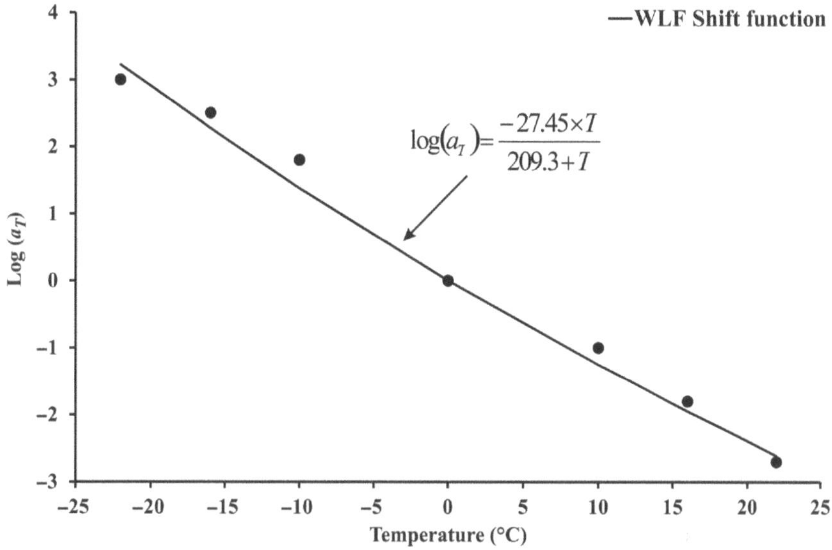

Figure 5.6 Shift factors and WLF shift function for constructing the master curve of complex shear modulus of an original bitumen at a reference temperature of 0°C.

Then, a_T can be determined which relates reduced frequency (or reduced time) to real frequency (or real time) as follows:

$$a_T = \frac{\omega_R}{\omega} \tag{5.3}$$

where ω_R is the reduced frequency which appears in the master curve and ω is the real frequency in the experiment, and

$$a_T = \frac{t}{t_R} \tag{5.4}$$

where t_R is the reduced time in the master curve and t is the real time of the experiment. Thus a_T for this example is as follows:

$$a_T = (10)^{-0.765} = 0.1718 = \frac{\omega_R}{\omega} \tag{5.5}$$

Calculating Eq. (5.5) for $\omega = 0.1$, and $\omega = 100$ shows that reduced frequencies corresponding to these real frequencies are $\omega_R = 0.01718$ and $\omega_R = 17.18$. Therefore it is possible to determine G^* for $\omega = 0.1$, and $\omega = 100$ at test temperature of $T_a = 6°C$ by capturing G^* of $\omega_R = 0.01718$ and $\omega_R = 17.18$ on the 0°C master curve. This simple example shows the power of TTSP to predict a desired viscoelastic

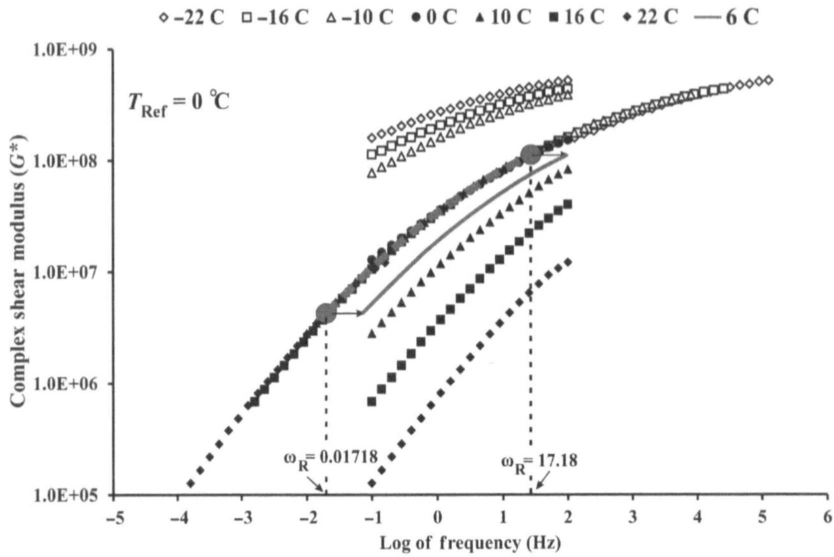

Figure 5.7 Procedure of determination complex shear modulus (G^*) at an arbitrary temperature by using the 0°C master curve and WLF shift function.

function at any arbitrary temperature by performing limited experiments at other test temperatures. Fig. 5.7 shows this procedure to better understand how it is possible to determine a viscoelastic function at a desired temperature having a master curve at a reference temperature and a shift function.

5.4.2 Arrhenius activation energy equation

The WLF equation is known to be only valid for temperatures greater than glass transition temperature, T_g, since for temperatures lower than T_g, it is not possible to consider a viscoelastic material as a supercooled liquid (Brinson & Brinson, 2008). Ferry (1980) pointed out that the slope of shift factor function should be discontinuous at the T_g as the thermal expansion coefficient suffers a discontinuity at the T_g.

The Arrhenius activation energy equation (Menzinger & Wolfgang, 1969) shown in Eq. (5.6) was used to develop a shift function for temperatures below the T_g.

$$\tau(T) = A\exp\left(-\frac{E_a}{RT}\right) \tag{5.6}$$

where τ the is relaxation time, T is the absolute temperature, E_a is the activation energy, and R is the gas constant. Applying the ln operation on both sides of Eq. (5.6) resulted in:

$$\ln[\tau(T)] = \ln(A) - \frac{E_a}{RT} \tag{5.7}$$

Then, taking the ratio at an arbitrary temperature, the T_g will be given after converting to base 10 logarithm as:

$$\log(a_T) = \log\left(\frac{\tau(T)}{\tau(T_g)}\right) = -\frac{E_a}{2.303R}\left(\frac{1}{T} - \frac{1}{T_g}\right) \tag{5.8}$$

It is necessary to mention that it is possible to substitute T_g in Eq. (5.8) with T_{ref} to use this equation for constructing a master curve for temperatures below glass transition temperatures.

5.5 Mathematical development of the time–temperature superposition principle

To organize the mathematical procedure of constructing a master curve by implementing time–temperature superposition principle (TTSP), it is possible to use the relaxation spectra which are defined in the form of the following equation:

$$E(t) = \int_0^{\infty} H(\tau)\exp\left(-\frac{t}{\tau}\right)d\tau \tag{5.9}$$

The relaxation spectra are a continuous form of a generalized Maxwell model (see Fig. 3.12), which is explained by using Prony series as Eq. (3.25) which is repeated here:

$$E(t) = \sum_{i=1}^{n} E_i\exp\left(-\frac{t}{\tau_i}\right) \tag{5.10}$$

It is evident that Eq. (5.10) is composed of discrete values of relaxation times. Therefore by increasing the number of elements in the generalized Maxwell element, it is possible to write the following equation:

$$E(t) = \lim_{n\to\infty} \sum_{i=1}^{n} E_i\exp\left(-\frac{t}{\tau_i}\right) \tag{5.11}$$

Multiplying and dividing the right-hand side of Eq. (5.11) by $\Delta\tau_i$ (an increment of τ_i) results in the following:

$$E(t) = \lim_{\substack{n \to \infty \\ \Delta\tau_i \to 0}} \sum_{i=1}^{n} \frac{E_i}{\Delta\tau_i} \exp\left(-\frac{t}{\tau_i}\right) \Delta\tau_i \tag{5.12}$$

where $E_i/\Delta\tau_i$ is similar to a singularity function (or a Dirac delta) and can be defined as $H(\tau_i)$. Therefore Eq. (5.12) can be represented in an integration form of Eq. (5.9) in which $H(\tau)$ is a continuous function entitled relaxation spectrum.

Assuming the generalized Kelvin–Voigt model and using a similar procedure on Eq. (3.24), a retardation spectrum can also be developed as:

$$D(t) = \int_0^{\infty} L(\tau)\left[1 - \exp\left(-\frac{t}{\tau}\right)\right] d\tau \tag{5.13}$$

In Eq. (5.9) or Eq. (5.13), it is evident that $H(\tau)$ or $L(\tau)$ are functions of temperature and these equations can be rewritten as:

$$E(t, T) = \int_0^{\infty} H(\tau(T), T)\exp\left(-\frac{t}{\tau}\right) d\tau \tag{5.14}$$

$$D(t, T) = \int_0^{\infty} L(\tau(T), T)\left[1 - \exp\left(-\frac{t}{\tau}\right)\right] d\tau \tag{5.15}$$

Passaglia and Knox noted that the explicit dependence of $H(\tau)$ on temperature is so weak and suggested to explain Eq. (5.14) as follows:

$$E(t, T) = \lambda T \int_0^{\infty} h(\tau)\exp\left(-\frac{t}{\tau}\right) d\tau \tag{5.16}$$

to simply relate E to T by using the parameter of λ. Therefore the $h(\tau)$ has no direct dependence on temperature. By multiplying both sides of Eq. (5.16) to $(\lambda_0 T_0/\lambda T)$:

$$\left(\frac{\lambda_0 T_0}{\lambda T}\right) E(t, T) = \lambda_0 T_0 \int_0^{\infty} h(\tau)\exp\left(-\frac{t}{\tau}\right) d\tau \tag{5.17}$$

Since Eq. (5.16) is valid for any temperature, it is possible to write this equation at T_0 as:

$$E(t, T_0) = \lambda_0 T_0 \int_0^{\infty} h(\tau_0)\exp\left(-\frac{t}{\tau_0}\right) d\tau \tag{5.18}$$

in which [similar to Eq. (5.4)], the relaxation time at T_0 can be related to the relaxation time at an arbitrary temperature of T by using a shift function of a_T as follows:

$$a_T = \frac{\tau}{\tau_0} \tag{5.19}$$

Having Eq. (5.19) and substituting τ_0 as τ/a_T in Eq. (5.18), this equation can be simply rewritten by defining a new time scale of t' as:

$$E(t', T_0) = \lambda_0 T_0 \int_0^{\infty} h(\tau)\exp\left(-\frac{t'}{\tau}\right) d\tau \tag{5.20}$$

where

$$t' = \frac{t}{a_T} \tag{5.21}$$

Comparing the right-hand side of Eq. (5.17) and Eq. (5.20) shows that both equations are identical. Therefore the following equations can be written for their left-hand sides:

$$\begin{aligned} E(t', T_0) &= \left(\frac{\lambda_0 T_0}{\lambda T}\right) E(t = a_T t', T) \\ E(t, T) &= \left(\frac{\lambda T}{\lambda_0 T_0}\right) E\left(t' = t/a_T, T_0\right) \end{aligned} \tag{5.22}$$

which clearly shows the TTSP in a mathematical statement. Eq. (5.22) shows that it is possible to determine the relaxation modulus at any temperature of T by multiplying a shift factor after modifying the timescale. In the literature, the new timescale (t') is sometimes referred to as "reduced time" as it is required to have vertical and horizontal shifts because of temperature differences in the experimental results. It is obvious that similar equations can be developed for the creep compliance. Using a log–log scale curve for sketching relaxation modulus versus time, the shift function can be represented as follows by applying a logarithm function on both sides of Eq. (5.21):

$$\log[t] = \log[t'] + \log[a_T] \tag{5.23}$$

in which a_T can be any type of shift functions. Fig. 5.8 shows the shifting procedure based on the abovementioned equations.

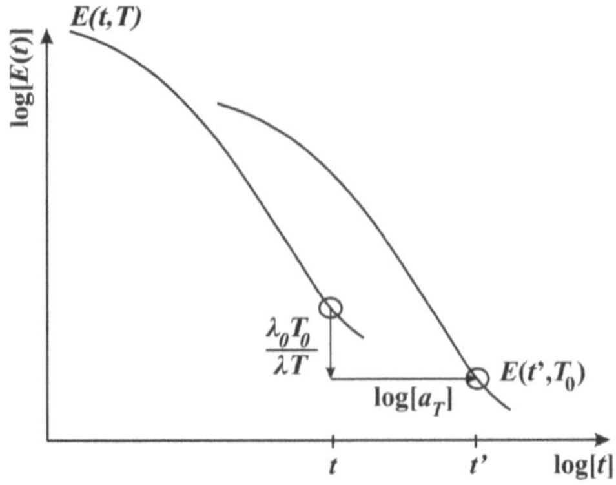

Figure 5.8 Vertical and horizontal shifting of relaxation modulus data for $T_0 < T$.

It is necessary to mention that in many cases, the change in density (λ) and the amount of change due to temperature (T) can be neglected because of their small values. Therefore the TTSP equation is simply written as:

$$E(t', T_0) = E(t = a_T t', T) \tag{5.24}$$

Nevertheless, neglecting vertical shift, even if small, can result in considerable error when predicting viscoelastic properties over a long time. Brinson and Brinson (2008) explained this type of error by presenting a master curve for the creep compliance of a polymer, as depicted in Fig. 5.9. As can be seen in this figure, if the original data for $T = T_1$ just shifted horizontally by neglecting the vertical shift, a considerable error will result in both the creep compliance and the timescale. This error is cumulative and may be so large for a long time even if the temperature ratio is small. Thus it is required to precisely evaluate the effect of neglecting vertical shift on the accuracy of the output master curve.

5.6 Mathematical-based master curve construction methods

It is essential to use precise methods for constructing a master curve, based on experimental results, which represents an appropriate mathematical function. This function can be used for developing numerical frameworks as well as the manipulation of data. In this section, two mathematical-based methods are introduced to construct master curves: the sign control method and the method proposed by Chailleux et al. (2006).

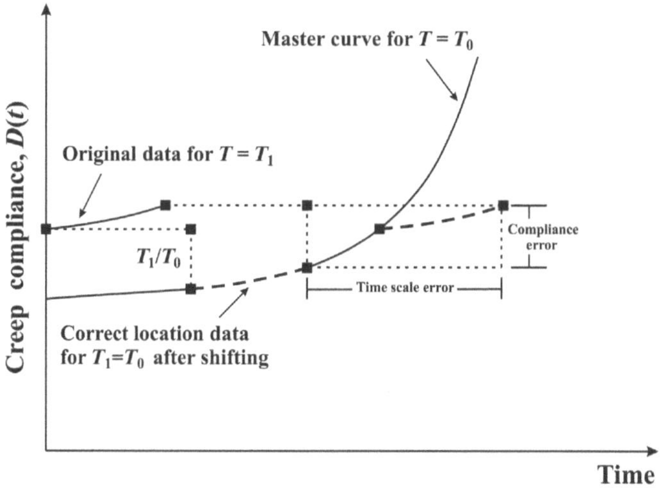

Figure 5.9 Probable error for lack of vertical shift (an example for creep compliance).

The following exponential Prony series was represented in Section 3.2.4 to describe the relaxation modulus of the generalized Maxwell model as well as the creep compliance of the generalized Kelvin–Voigt model:

$$E(t) = E_\infty + \sum_{i=1}^{n} E_i \exp\left(-\frac{t}{\tau_i}\right) \tag{5.25}$$

$$D(t) = D_g + \sum_{i=1}^{n} D_i\left(1 - \exp\left(-\frac{t}{\tau_i}\right)\right) \tag{5.26}$$

This form of representation of constitutive viscoelastic functions has three main advantages: (1) the coefficients of E_∞, D_g, E_i, D_i, and τ_i can simply be related to simple components of spring and dashpot as depicted in Figs. 3.11 and 3.12; (2) the derivatives and integration of the exponential function are simple so that it will be easy to manipulate the exponential terms mathematically; and (3) using the Prony series for developing a numerical framework to solve a boundary value problem enables the use of a recursive algorithm (Zhang & Li, 2009; Zhang et al., 2010) to determine the residual force and residual stress, which is so easy to implement, and will be discussed more in Chapter 8, Incremental Form of Viscoelastic Relations.

To construct a master curve at a reference temperature on experimental curves by considering Eqs. (5.25) and (5.26), the main challenge is to determine E_i and τ_i for the relaxation modulus and D_i and τ_i for the creep compliance to provide a proper fit over all time of the data. Several researchers have proposed a method for solving this problem: the collocation method by Schapery, the multidata method by Cost and Becker (1970), the windowing method by Emri and Tschoegl (1993), and the sign control method proposed by Bradshaw and Brinson.

5.6.1 Sign control method

The sign control method is actually based on the multidata method. The first step in this method is to select the relaxation times in a rational manner for mathematical convenience to proceed with constructing the desired master curve. Since a simple thermorheological viscoelastic material has a continuous master curve of relaxation modulus or creep compliance over time, it is essential to select an ample number of relaxation times that differ depending on the smoothness of the experimental data. The data smoothness is directly dependent on the number of test temperatures. Generally, 8–20 relaxation times over 10 decades of time can be an appropriate rule of thumb. Fig. 3.13 depicts an interesting representation about the importance of selecting an appropriate number of relaxation times. As can be seen in this figure, using two or five Maxwell models resulted in inappropriate fitting of Prony series function.

After selecting appropriate relaxation times, the number of unknown coefficients is halved and it is just required to find accurate E_i or D_i such that the Prony series of the relaxation modulus or the creep compliance optimally fits on the provided time domain data. For this purpose, a generalized least square approach can be used in which the coefficients are determined that minimize the following error function between the observed data and the calculated data.

$$\Lambda^2 = \sum_{k=1}^{K} \left(\frac{E(t_k) - E_k}{\sigma_k} \right)^2 \tag{5.27}$$

in which $E(t_k)$ is the calculated data based on Prony series function at time t_k, E_k is the experimental data [K data pairs as (E_k, t_k)], and σ_k is the standard deviation of the kth data point. Without defining a constraint to minimize the Eq. (5.27), it is probable to capture positive and negative values for E_i. However, with regard to the physical meaning of the stiffness of a spring, it is necessary to define a constraint of $E_i > 0$. It is apparent that such a formulation can be used to calculate D_i for the creep compliance data. Furthermore, this procedure can be implemented to fit frequency domain data where experimental measurements were done by using, for example, a dynamic mechanical analyzer (DMA) or dynamic shear rheometer (DSR) which is further explained in the next chapter.

5.6.2 Chailleux's method

Chailleux et al. (2006) presented a mathematical-based procedure to construct master curves of a complex modulus and phase angle based on Kramers–Kronig relations which are as follows:

$$\log\left|G*(\omega)\right| - \log\left|G*(\infty)\right| = -\frac{2}{\pi}\int_0^\infty \frac{u\delta(u) - \omega\delta(\omega)}{u^2 - \omega^2}du \tag{5.28}$$

$$\delta(\omega) = \frac{2\omega}{\pi}\int_0^\infty \frac{\log\left|G*(u)\right| - \log\left|G*(\omega)\right|}{u^2 - \omega^2}du \tag{5.29}$$

By simplifying Eq. (5.28), Booij and Thoone (1982) presented the following approximation:

$$\delta(\omega) \approx \frac{\pi}{2}\frac{d\log\left[\left|G*(\omega)\right|\right]}{d\log(\omega)} \tag{5.30}$$

which will be exact when material behavior can be represented by using a general power law of $f(t) = At^\alpha$ as introduced in Eq. (3.50). Therefore Eq. (5.30) can be rewritten as:

$$\delta(\omega) = \frac{\pi}{2}\frac{d\log\left[\left|G*(\omega)\right|\right]}{d\log(\omega)} \tag{5.31}$$

In order to determine the shift function, the Eq. (5.31) can be used by considering two close frequencies of ω_i and ω_j where $\omega = 2\pi f$. The following equation can be written by using the basic definition of the derivation:

$$\delta_{average}^{(\omega_i,\, \omega_j)}\frac{2}{\pi} = \frac{\log\left(\left|G*(T,\, \omega_j)\right|\right) - \log\left(\left|G*(T,\, \omega_i)\right|\right)}{\log(\omega_j) - \log(\omega_i)} \tag{5.32}$$

in which δ_{average} is the average of two phase angles measured at test temperature of T at frequencies of ω_i and ω_j. It is also possible to perform an interpolation of phase angle based on test temperature to better approximate it. Considering the TTSP and a shift factor for relating test temperature of T_1 and T_2 as $a_{(T1,\ T2)} = f_2/f_1 = \omega_2/\omega_1$ such that $|G^*(T_1, \omega_1)| = |G^*(T_2, \omega_2)|$, it is possible to rewrite Eq. (5.32) as follows:

$$\delta_{average}^{(T_1,\ T_2)}(\omega_2)\frac{2}{\pi} = \frac{\log(|G*(T_1,\ \omega_2)|) - \log(|G*(T_2,\ \omega_2)|)}{\log(a_{(T_1,T_2)})} \tag{5.33}$$

where δ_{average} is the average of two phase angles measured at two different test temperatures of T_1 and T_2 at frequency of ω_2. Thus, shift factors can be determined by using Eq. (5.33) for close isotherms. If we have measurements at different temperatures of T_1, T_2, ..., T_n, it will be possible to construct a master curve at a reference temperature of T_{ref} ($1 < \text{ref} < n$) by applying a cumulative sum of log ($a_{(Ti,\ Ti+1)}$) as Eq. (5.34) shows such an application for an isotherm T_i based on the reference temperature T_{ref} as follows:

$$\log\left(a_{(T_i,\ T_{ref})}\right) = \sum_{j=i}^{j=ref} \log\left(a_{(T_j,\ T_{j+1})}\right) \tag{5.34}$$

and then:

$$\log\left(a_{(T_i,T_{ref})}\right) = \sum_{j=i}^{j=ref} \frac{\pi}{2} \frac{\log(|G*(T_j,\omega)|) - \log(|G*(T_{j+1},\omega)|)}{\delta_{average}^{(T_j,T_{j+1})}(\omega)} \tag{5.35}$$

After determining all required shift factors by using Eq. (5.35), it is possible to use the shift functions introduced in Section 5.4. For example, considering the WLF equation resulted in the following function:

$$\log\left(a_{(T_i,T_{ref})}\right) = \frac{-c_{1,\ ref} \times \left(T_i - T_{ref}\right)}{c_{2,\ ref} + \left(T_i - T_{ref}\right)} \tag{5.36}$$

In the case of constructing a master curve at a reference temperature (T'_{ref}) which is not measured in the laboratory, it is possible to rewrite Eq. (5.36) by using other c_1 and c_2 constant coefficients because the form of Eq. (5.36) is independent of the choice of T_{ref} (Ferry, 1980). The relation between $c_1{'}_{\text{ref}}$ and $c_2{'}_{\text{ref}}$ is:

$$c_{2,\ ref} = c'_{2,\ ref} + T_{ref} - T'_{ref} \tag{5.37}$$

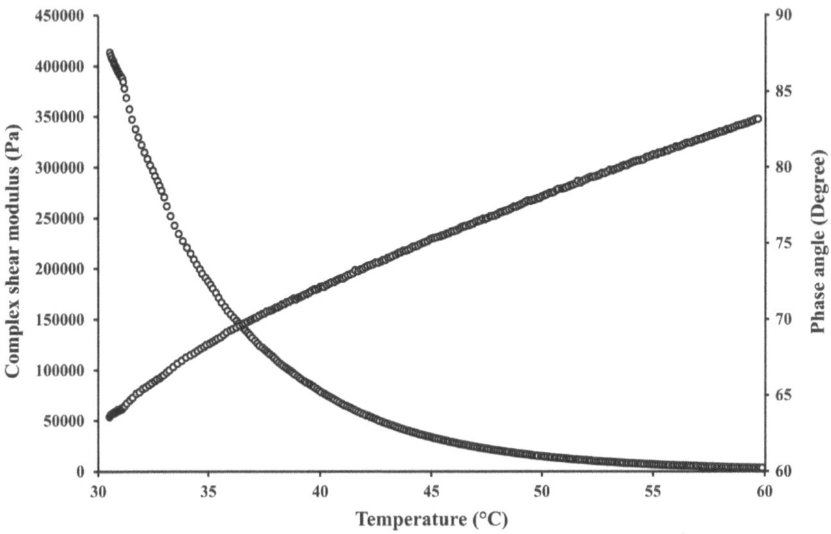

Figure 5.10 Results of temperature sweep test for an original bitumen at frequency of 1.59 Hz.

and

$$c_{1,ref} = \frac{c'_{1,ref} \times c'_{2,ref}}{c_{2,ref}} \tag{5.38}$$

Therefore having $c_{1,\text{ref}}$ and $c_{2,\text{ref}}$ at T_{ref}, it will be simply possible to construct master curves at any temperature.

As an example for applying the presented method on experimental data, Fig. 5.10 shows results (both complex shear modulus and phase angle) of a temperature sweep test of an original bitumen in which the test temperature sweeps from 30°C to 60°C and the experiment was performed at a frequency of 1.59 Hz. It is necessary to mention that the temperature sweep test is usually performed to observe the rheological behavior of a viscoelastic material within a temperature range at a specific frequency.

By using the abovementioned procedure, the master curves of complex shear modulus (G*) and phase angle (δ) are constructed at the reference temperature of 45°C and shown in Fig. 5.11. Fig. 5.12 also depicts a shift factor for this master curve determined by implementing Eq. (5.35). As can be seen in Fig. 5.11, performing a temperature sweep at one frequency of 1.59 Hz, a master curve can be constructed in a wide

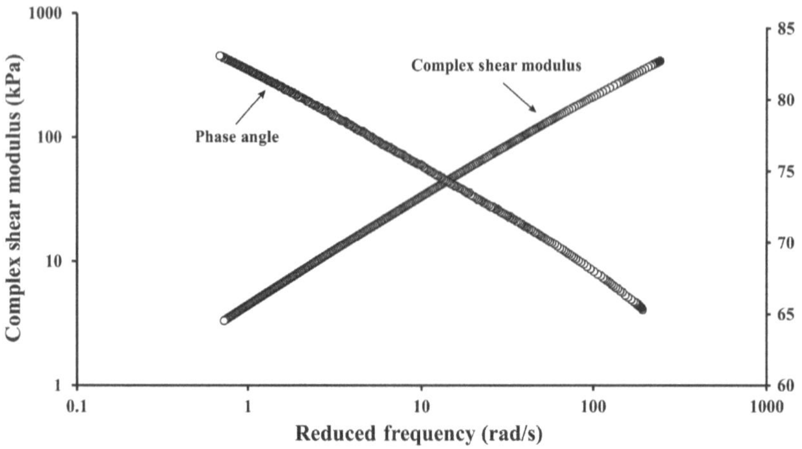

Figure 5.11 Master curves of (A) complex shear modulus and (B) phase angle at the reference temperature of 45°C.

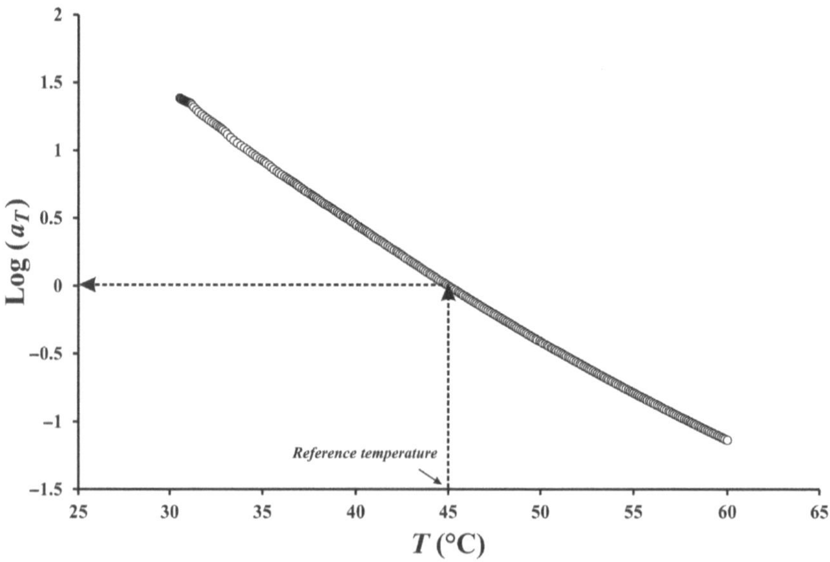

Figure 5.12 Shift factor versus temperature to construct master curve of complex shear modulus and phase angle of an original bitumen at the reference temperature of 45°C.

range of frequency. It is evident that it will be possible to fit the WLF function on the shift factor curve represented in Fig. 5.12 to determine c_1 and c_2.

5.7 Constitutive equations with effective time

In Section 2.3, the following constitutive equations are derived for relating stress and strain based on Boltzmann's superposition principle:

$$\sigma(t) = \int_0^t E(t-\xi)\frac{d\varepsilon(\xi)}{d\xi}d\xi \tag{5.39}$$

$$\varepsilon(t) = \int_0^t D(t-\xi)\frac{d\sigma(\xi)}{d\xi}d\xi \tag{5.40}$$

in which $\sigma(t)$ is stress, $\varepsilon(t)$ is strain, $E(t)$ is the relaxation modulus, $D(t)$ is the creep compliance, ξ is auxiliary parameter of integration, and t is time. The constitutive viscoelastic functions including the relaxation modulus and the creep compliance are dependent on temperature. Therefore it is essential to be able to account for such an effect in these equations. Considering the TTSP, the relaxation modulus at an arbitrary temperature of T can relate to the relaxation modulus at the reference temperature of T_{ref} by implementing the shift factor a_T as:

$$E(t, T) = E\left(\zeta = t/a_T, T_{ref}\right) \tag{5.41}$$

in which ζ is reduced time. Considering a small increment of dt, the relaxation modulus at temperature T has changed to $E(t + dt)$ and accordingly the relaxation modulus at the reference temperature will be $E(\zeta + d\zeta)$ such that:

$$E(t + dt, T) = E\left(\zeta + d\zeta, T_{ref}\right) \tag{5.42}$$

in which, it is obvious that:

$$d\zeta = \frac{dt}{a_T(T(t))} \tag{5.43}$$

Eq. (5.43) explains that all relaxation times at temperature T are $1/a$ times faster or slower than those taking place in the reduced time increment at the reference temperature. The effective time is defined as the integration of Eq. (5.43):

$$\zeta(t) = \int_0^t \frac{d\psi}{a_T(T(\psi))} \tag{5.44}$$

It is essential to involve temperature history to determine the stress and strain based on the constitutive equation of Eqs. (5.39) and

(5.40). For example, Eq. (5.39) can be written in the effective time domain as:

$$\sigma(\zeta) = \int_0^{\zeta} E(\zeta - \zeta', T_{ref}) \frac{d\varepsilon(\zeta')}{d\zeta'} d\zeta' \tag{5.45}$$

and then it is possible to map the abovementioned equation to the real-time domain as follows:

$$\sigma(t) = \int_0^{t} E(\zeta(t) - \zeta(t'), T_{ref}) \frac{d\varepsilon(t')}{dt'} dt' \tag{5.46}$$

where

$$\zeta(t) = \int_0^{t} \frac{d\psi}{a_T(T(\psi))} \tag{5.47}$$

and

$$\zeta(t') = \int_0^{t'} \frac{d\psi}{a_T(T(\psi))} \tag{5.48}$$

Eq. (5.46) can be simply used in problems with known temperature and strain as functions of time. However, since this equation is not a convolution integral and is difficult to solve, often problems are solved in the effective time domain by using Eq. (5.45). Using a similar procedure, the following equation can be developed for the strain as:

$$\varepsilon(t) = \int_0^{t} D(\zeta(t) - \zeta(t'), T_{ref}) \frac{d\sigma(t')}{dt'} dt' \tag{5.49}$$

References

Booij, H. C., & Thoone, G. P. J. M. (1982). Generalization of Kramers-Kronig transforms and some approximations of relations between viscoelastic quantities. *Rheologica Acta, 21*(1), 15–24. Available from https://doi.org/10.1007/BF01520701.

Brinson, H. F., & Brinson, L. C. (2008). *Polymer engineering science and viscoelasticity: An introduction*. Springer.

Chailleux, E., Ramond, G., Such, C., & de La Roche, C. (2006). A mathematical-based master-curve construction method applied to complex modulus of bituminous materials. *Road Materials and Pavement Design*, 7(Suppl. 1), 75–92. Available from https://doi.org/10.1080/14680629.2006.9690059.

Cost, T. L., & Becker, E. B. (1970). A multidata method of approximate Laplace transform inversion. *International Journal for Numerical Methods in Engineering*, *2*(2), 207–219. Available from https://doi.org/10.1002/nme.1620020206.

Emri, I., & Tschoegl, N. W. (1993). Generating line spectra from experimental responses. Part I: Relaxation modulus and creep compliance. *Rheologica Acta, 32*(3), 311–322. Available from https://doi.org/10.1007/BF00434195.

Ferry, J. D. (1980). *Viscoelastic properties of polymers*. John Wiley & Sons.

Menzinger, M., & Wolfgang, R. (1969). The meaning and use of the Arrhenius activation energy. *Angewandte Chemie International Edition in English, 8*(6), 438–444. Available from https://doi.org/10.1002/anie.196904381.

Pelleg, J. (2012). *Mechanical properties of materials* (Vol. 190). Springer Science & Business Media.

Sadd, M. H. (2009). *Elasticity: Theory, applications, and numerics*. Academic Press.

Tschoegl, N. W. (2012). *The phenomenological theory of linear viscoelastic behavior: An introduction*. Springer Science & Business Media.

Williams, M. L., Landel, R. F., & Ferry, J. D. (1955). The temperature dependence of relaxation mechanisms in amorphous polymers and other glass-forming liquids. *Journal of the American Chemical Society*, 77(14), 3701–3707.

Zhang, H. H., & Li, L. X. (2009). Modeling inclusion problems in viscoelastic materials with the extended finite element method. *Finite Elements in Analysis and Design, 45* (10), 721–729. Available from https://doi.org/10.1016/j.finel.2009.06.006.

Zhang, H. H., Rong, G., & Li, L. X. (2010). Numerical study on deformations in a cracked viscoelastic body with the extended finite element method. *Engineering Analysis with Boundary Elements, 34*(6), 619–624. Available from https://doi.org/10.1016/j.enganabound.2010.02.001.

CHAPTER 6

Interconversion of constitutive viscoelastic functions

6.1 Introduction

As it is mentioned in Chapter 2, Constitutive Viscoelastic Functions, there are two constitutive functions for characterizing the rheological and mechanical behavior of viscoelastic materials: relaxation modulus and creep compliance. The creep compliance is defined as the ratio of time-dependent strain to time-dependent stress, as shown in Eq. (2.1), and the relaxation modulus is the ratio of time-dependent stress to time-dependent strain, as shown in Eq. (2.4). For mechanical modeling of creep and relaxation behavior of viscoelastic materials, such as asphalt binder, asphalt mastic, asphalt mortar, and asphalt mixture, spring and dashpot components are used, which characterize the elastic and viscous properties, respectively (Huang, 1993). Many of the linear viscoelastic models are created by the combination of the two abovementioned components when the differential equation approach is selected for describing the mechanical behavior of viscoelastic materials. The Maxwell model, the Kelvin–Voigt model, the Burgers model, and the generalized Maxwell of Kelvin models are typical linear viscoelastic constitutive models as introduced in Chapter 3, Mechanical Models of Viscoelasticity (see Section 3.2). It is necessary to mention that these models must be capable of generating both the relaxation modulus and the creep compliance accurately based on experimental data or analytical solution.

The main challenge for the characterization of viscoelastic materials is to design and perform experiments which are able to determine these constitutive functions precisely. Regarding the nature of the desired viscoelastic material and depending on the temperature and loading conditions, it is probable that measuring one of these two constitutive functions is not possible, accurate, or may be so expensive. For example, controlling a constant strain for capturing relaxation modulus is commonly more challenging than controlling a constant stress. Therefore stress-control experiments are more common in comparison with strain-control ones. To overcome such a

Applications of Viscoelasticity
DOI: https://doi.org/10.1016/B978-0-12-821210-3.00005-X

shortcoming for modeling viscoelastic materials, it is well-known that by having one of these two functions, another one can be calculated considering the general relation between the relaxation modulus and the creep compliance in the Laplace-transformed domain based on the correspondence principle (CP) (see Chapter 4: Correspondence Principle of Viscoelasticity) (Fabrizio & Morro, 1992; Phan-Thien & Rossikhin, 2004). Several research works have been devoted to introducing simple and efficient methods for performing such an interconversion to determine an unknown viscoelastic function based on other measured and known functions.

Hopkins and Hamming (1957) introduced a general concept to interconvert all kinds of creep and relaxation viscoelastic functions based on dividing a specific problem into several time steps and using some simplifying assumptions to determine the unknown function. Mead (1994) modeled the relaxation modulus as a discrete N element Maxwell ($N \gg 1$) line spectrum. The method incorporates additional independent rheological data into a constrained linear regression with regularization. Kim et al. (1997) implemented a method based on the CP for fatigue characterization of asphalt concrete, introduced earlier for the characterization of asphalt concrete (Kim & Lee, 1995; Kim et al., 1995). Park and Schapery (1999) proposed a numerical method of interconversion between linear viscoelastic material functions based on the Prony series representation. Park and Kim (1999) introduced a new approximate interconversion scheme using the local log–log slope of the source function. They developed this model by using a mathematical relationship based on the power-law representations of relaxation modulus and creep compliance. Also, Sorvari and Malinen (2007) implemented the numerical evaluation of the convolution integral to studying the interconversion between linear viscoelastic material functions, emphasizing materials with a relatively fast rate of relaxation. To combine the outputs of bending beam rheometer (BBR) and dynamic shear rheometer (DSR) experimental tests in the frequency domain, Underwood et al. (2010) implemented the general power law to express the calculated creep compliance function according to the mid-span beam deflections measured during the BBR test. Recently, inspiring the idea of using the power law to describe the constitutive linear viscoelastic functions of bituminous materials, Hajikarimi et al. (2018) implemented the general power law to interconvert linear viscoelastic functions of modified asphalt binders focusing on the determination of relaxation modulus to be used in a finite element numerical framework. They showed that it is possible to simply determine relaxation modulus of asphalt binders at low temperatures by using creep compliance data and considering a simple fractional viscoelastic model (Fini et al., 2016).

In this chapter, different well-known and practical interconversion methods are introduced, including Hopkins and Hamming's method, the system of algebraic equations method based on Prony series, general approximation methods of interconversion, and an approximation method based on the general power law. For each method, some applicable examples are introduced based on analytical solutions or experimental data. Additionally, MATLAB® codes are provided for performing the solution procedure.

6.2 Hopkins and Hamming's method

6.2.1 Interconverting relaxation modulus and creep compliance

As was mentioned in Section 4.2, the fundamental principle of the relationship between the relaxation modulus, $E(t)$, and the creep compliance, $D(t)$, for linear viscoelastic materials in the convolution integral form can be shown as Eq. (6.1) or equivalently as Eq. (6.2) (Ottosen & Ristinmaa, 2005):

$$\int_0^t E(\xi)D(t-\xi)d\xi = t \tag{6.1}$$

$$\int_0^t E(t-\xi)D(\xi)d\xi = t \tag{6.2}$$

in which ξ is the auxiliary parameter of integration. Since $E(t) \neq 1/D(t)$ [see Eq. (4.22)], there is no straightforward procedure for determining the relaxation modulus based on the measured creep compliance and vice versa. Hopkins and Hamming (1957) proposed a mathematical procedure to determine the creep compliance and the relaxation modulus by solving Eqs. (6.1) and (6.2) as presented in the following or other functions like bulk modulus and Poisson's ratio for corresponding viscoelastic materials as will be presented in this section. For instance, it is assumed that relaxation modulus data are available and aimed to calculate the creep compliance. In other words, $E(t)$ is measured and is known as raw data or a fitted curve, while $D(t)$ is the unknown desired material property. For this purpose, the convolution integral represented in Eq. (6.2) is discretized into n time steps as follows:

$$t_n = \sum_{i=1}^{n} \int_{t_{i-1}}^{t_i} E(t_n - \xi)D(\xi)d\xi \tag{6.3}$$

By considering the trapezoidal rule for calculating the integral for each time step in Eq. (6.3), it can be derived as:

$$t_n = \sum_{i=1}^{n} D(t_{i-\frac{1}{2}}) \frac{E(t_n - t_i) + E(t_n - t_{i-1})}{2} [t_i - t_{i-1}] \tag{6.4}$$

where $\Delta t = [t_i - t_{i-1}]$ is assumed time step and $D(t_{i-1/2})$ can be assumed as one of the following equations:

$$D(t_{i-\frac{1}{2}}) = D\left(\frac{t_i + t_{i-1}}{2}\right) \tag{6.5}$$

$$D(t_{i-\frac{1}{2}}) = \frac{D(t_i) + D(t_{i-1})}{2} \tag{6.6}$$

For better understanding, both the abovementioned assumptions are illustrated in Fig. 6.1 and it is evident that selecting Eqs. (6.5) or (6.6) significantly depends on the form of the experimental data or fitted curve.

For the first step ($n = 1$), the following equation can simply be found by substituting $n = 1$ in the Eq. (6.4):

$$t_1 = D(t_{1/2}) \frac{E(0) + E(t_1)}{2} t_1 \Rightarrow D(t_{1/2}) = \frac{2}{E(0) + E(t_1)} \tag{6.7}$$

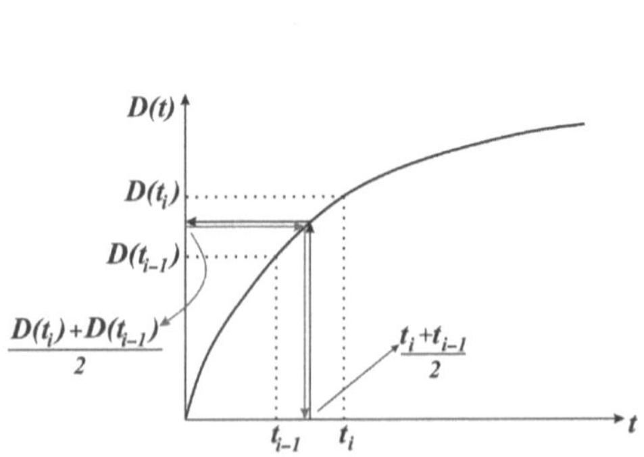

Figure 6.1 Approximation of $D(t)$ for a time step in the Hopkins and Hamming's method.

To continue the procedure of solving, the last term is removed from the summation as follows:

$$t_n = \sum_{i=1}^{n-1} D(t_{i-\frac{1}{2}}) \frac{E(t_n - t_i) + E(t_n - t_{i-1})}{2}[t_i - t_{i-1}] + D(t_{n-\frac{1}{2}}) \frac{E_g + E(t_n - t_{n-1})}{2}[t_n - t_{n-1}] \quad (6.8)$$

in which E_g is glassy modulus at $t = 0$ and $D(t_{i-1/2})$ can simply be determined as a recursive function of previous terms as:

$$D(t_{n-\frac{1}{2}}) = \frac{2}{\left[E_g + E(t_n - t_{n-1})\right]\Delta t_n} \left(t_n - \sum_{i=1}^{n-1} D(t_{i-\frac{1}{2}}) \frac{E(t_n - t_i) + E(t_n - t_{i-1})}{2} \right) \Delta t_i \quad (6.9)$$

where $\Delta t_i = t_i - t_{i-1}$.

By applying a similar procedure for determining the relaxation modulus based on creep compliance data, the following equation is derived:

$$E(t_{n+\frac{1}{2}}) = \frac{t_{n+1} - \sum_{i=0}^{n-1} E(t_{i+\frac{1}{2}})\left[f(t_{n+1} - t_i) - f(t_{n+1} - t_{i+1})\right]}{f(t_{n+1} - t_n)} \quad (6.10)$$

where $E(t)$ and t are the relaxation modulus and time, respectively, and the f function can be represented as the following recursive equation:

$$f(t_{n+1}) = f(t_n) + \frac{1}{2}[D(t_{n+1}) + D(t_n)][t_{n+1} - t_n] \quad (6.11)$$

in which $D(t)$ is the creep compliance. With regard to Eq. (6.4), $f(t = 0)$, the initial value of f function is considered to be equal to zero.

To provide a numerical framework to use the Hopkins and Hamming's method, Box 6.1 presents a simple MATLAB code to determine the relaxation modulus based on creep compliance data.

To show the ability of the MATLAB code presented in Box 6.1, the results of a BBR was considered to determine the relaxation modulus of an asphalt binder. According to ASTM D6648, the creep compliance was measured during a three-point bending beam test (see Section 4.3.1) for 240 s in which the mid-span deflection is captured every 0.5 s. Fig. 6.2 shows the creep compliance curve as well as the relaxation modulus curve. Having discrete relaxation modulus in 0.5 s time steps, it will be possible to fit an arbitrary function with regard to the mechanical model which is selected for describing the continuous viscoelastic behavior of the

BOX 6.1 MATLAB code for determination of relaxation modulus based on creep compliance data by using Hopkins and Hamming's method.

```
%Finding f function

%numpts is the number of experimental data points of the creep %
compliance

f = zeros(numpts,1);
f(1)= 0;
for i =1:numpts
    f(i+1)=f(i)+0.5*(D(i+1)+D(i)).*(t(i+1)-t(i));
end

%Determination of relaxation modulus E(t)
E = zeros(numpts,1);
Sigma = 0;
B = [];
for i = 0:numpts
    sigma = zeros(i,1);

        for n = 1:i
            sigma(n) = E(i-n+1)*(f(n+2)-f(n+1));
        end

    B = sum(sigma);
    E(i+1)=(t(i+1)-B)/f(2);
End
```

material. On the other hand, it is possible to use discrete data of the relaxation modulus for numerical simulation.

6.2.2 Derivation of bulk modulus based on relaxation modulus and Poisson's ratio

Another example of using Hopkins and Hamming's method is presented here where the extension relaxation modulus, $E(t)$, and the Poisson's ratio as a function of time, $\nu(t)$ is known and the bulk modulus, $K(t)$ is desired. There is an elastic relation to calculate the bulk modulus having the Young's modulus and the Poisson's ratio as (Sadd, 2009):

$$K = \frac{E}{3(1 - 2\nu)} \tag{6.12}$$

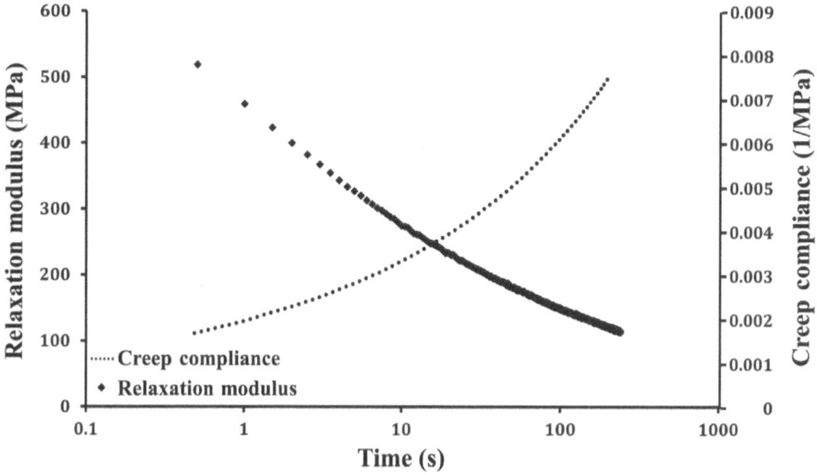

Figure 6.2 Measure creep compliance and determined relaxation modulus curve of a BBR test.

At the first step, by using the CP (see Chapter 4: Correspondence Principle of Viscoelasticity), it is possible to substitute all variables in Eq. (6.12) with their Laplace transforms as:

$$s\overline{K}(s) = \frac{s\overline{E}(s)}{3(1 - 2s\overline{\nu}(s))} \tag{6.13}$$

To proceed further, a simple trick can be used by adding and subtracting an extra term to Eq. (6.13) as follows:

$$\begin{aligned} \overline{E}(s) &= 3(1 - 2s\overline{\nu}(s))\overline{K}(s) + \left[6\nu_g\overline{K}(s) - 6\nu_g\overline{K}(s)\right] \\ &= 3\overline{K}(s) - 6\left(s\overline{\nu}(s) - \nu_g\right)\overline{K}(s) - 6\nu_g\overline{K}(s) \end{aligned} \tag{6.14}$$

in which ν_g is $\nu(t=0)$. Then, by applying the Laplace transform on Eq. (6.14) and remembering Eq. (4.6) for the Laplace transformation of a function's derivation, the following equation can be derived:

$$E(t) = \left(3 - 6\nu_g\right)K(t) - 6\int_0^t K(t-\xi)\frac{d\nu}{d\xi}d\xi \tag{6.15}$$

At the second step, the Hopkins and Hamming's method is used in a similar way as applied in Section 6.2.1. The total time $[0, t]$ is discretized into n time steps and accordingly, Eq. (6.15) can be written as follows:

$$E(t_n) = \left(3 - 6\nu_g\right)K(t_n) - 6\sum_{i=1}^{n}\int_{t_{i-1}}^{t_i} K(t_n-\xi)\frac{d\nu}{d\xi}d\xi \tag{6.16}$$

Then, by using the trapezoidal rule for integration, the integral in the right side of Eq. (6.16) can be simply calculated as:

$$K(t_n)(3-6\nu_g) = E(t_n) + 6\sum_{i=1}^{n}\frac{K(t_n-t_i)+K(t_n-t_{i-1})}{2}[\nu(t_i)-\nu(t_{i-1})] \tag{6.17}$$

The second part of the summation in the right-hand side of Eq. (6.17) can be written by the variable transformation of $j = i - 1$ as follows:

$$\sum_{i=1}^{n}K(t_n-t_{i-1})[\nu(t_i)-\nu(t_{i-1})] \xrightarrow{j=i-1} \sum_{j=0}^{n-1}K(t_n-t_j)[\nu(t_{j+1})-\nu(t_j)] \tag{6.18}$$

and then by substituting $j = i$ again, the following equation can be derived in which the summation starts from 0 to $n - 1$:

$$\sum_{i=0}^{n-1}K(t_n-t_i)[\nu(t_{i+1})-\nu(t_i)] \tag{6.19}$$

Now, having Eq. (6.19), it is possible to calculate Eq. (6.17) as:

$$\begin{aligned}K(t_n)(3-6\nu_g) = {} & E(t_n) + 6\sum_{i=1}^{n-1}\frac{K(t_n-t_i)}{2}[\nu(t_i)-\nu(t_{i-1})] \\ & + \frac{K(t_n-t_i)}{2}[\nu(t_{i+1})-\nu(t_i)] + \underbrace{6\frac{K(t_n-t_n)}{2}[\nu(t_n)-\nu(t_{n-1})]}_{\text{for } i=n} \\ & + \underbrace{6K(t_n)[\nu(t_1)-\nu_g]}_{\text{for } i=0}\end{aligned} \tag{6.20}$$

and finally, the equation can be condensed as:

$$K(t_n)(3-6\nu_g) = E(t_n) + 3K_g[\nu(t_n)-\nu(t_{n-1})] + 6\sum_{i=1}^{n-1}\frac{K(t_n-t_i)}{2}[\nu(t_{i+1})-\nu(t_{i-1})] \tag{6.21}$$

where K_g is $K(t=0)$. As can be seen in Eq. (6.21), determination of $K(t_n)$ needs one to have $K(t_n-t_i)$. For equal time steps from the beginning up to the end of the solution, there is no problem and Hopkins and Hamming's method can work properly. However, for nonequal time steps, it is not possible to use this method because the relaxation bulk modulus is unknown for each t_n-t_i.

It is necessary to mention that a similar procedure can be used for determining the shear modulus, G, by using the relaxation modulus, E, and the Poisson's ratio, ν, based on the following equation (Sadd, 2009):

$$G = \frac{E}{2(1+\nu)} \tag{6.22}$$

6.2.3 Derivation of Poisson's ratio based on complex shear modulus and relaxation modulus

Asphalt mixtures consist of asphalt binder, mineral aggregates, and often a filler material, and generally behave as a viscoelastic material, due to the presence of an asphalt binder. By definition, as a viscoelastic material, the response of asphalt mixtures is highly dependent on loading frequency (time) and temperature (Lee & Kim, 2009). To understand the mechanical behavior of an asphalt mixture, the stress–strain relationship has generally been used to account for essential asphalt mixture characteristics, such as creep compliance and relaxation modulus. Such characteristics are vital for flexible pavement design in terms of the pavement response prediction over time, under traffic and environmental loading. For flexible pavement design purposes, it is often assumed that the Poisson's ratio, ν, is time independent. However, this assumption is fundamentally incorrect due to the viscoelastic nature of asphalt mixtures, a property resulting from the inclusion of asphalt binder (Kassem et al., 2013). Often, the Poisson's ratio is simply assumed to be 0.35 (Lindelöf et al., 2018; Mallela et al., 2009) but can be determined from various methods. Witczak and Mirza (1999) suggested that asphalt mixture's Poisson's ratio can be formulated from indirect tensile test data. Zak et al. (2015) indicated that such values can be obtained using data acquired from small-amplitude oscillation tests.

Accurate determination of the Poisson's ratio of asphalt mixtures has long been a challenge. The resilient modulus test has often been used to determine the Poisson's ratio values (Zhang et al., 1997), although the practice has been criticized due to the presence of a biaxial stress pattern in the test (Di Benedetto et al., 2004); the test cannot be performed in a controlled strain mode. Rather than using imprecise methods to estimate asphalt mixture's ν values, or simply assume a constant asphalt mixture's ν, perhaps a more accurate asphalt mixture's Poisson's ratio value could be obtained using the asphalt binder's Poisson's ratio used in the mixture, since the binder is the primary source of mixture's viscoelastic properties.

Currently, there are no easily accomplished methods to accurately determine the Poisson's ratio values for asphalt binders. According to the mechanics of time-dependent materials (Tschoegl, 2012), it is possible to calculate material's Poisson's ratio values if the relaxation shear modulus, $G(t)$, and the relaxation modulus, $E(t)$, are available [see Eq. (6.22)]. However, applying this concept can be challenging. Conventionally, the relaxation shear modulus can be calculated from DSR data and is a function of frequency, $G(\omega)$, while the relaxation bending modulus is a function of loading time, $E(t)$, calculated using BBR data which are both thoroughly presented in Chapter 7, Application of Viscoelasticity for Experimental Tests. To calculate the Poisson's ratio for a binder, both moduli should be in the same domain, either frequency (ω) or time (t), making it necessary to convert one of the moduli. Given that DSR and BBR data are easily obtained in standard binder testing, the objective of this section is to provide an approach for changing relaxation shear modulus from the frequency to the time domain and then, combined with the relaxation bending modulus, determine the Poisson's ratio values for asphalt binders using a recursive numerical approach based on Hopkins and Hamming's method.

The main outputs of a DSR frequency sweep test are complex shear modulus (G^*) and phase angle (δ) which is further explained in Chapter 7, Application of Viscoelasticity for Experimental Tests. Fig. 6.3 depicts G^* and δ for an original asphalt binder at different test temperatures of 0°C, −6°C, −12°C, and −18°C.

By using the concept of time−temperature superposition introduced in Chapter 5, Time−Temperature Superposition, the master curve of the shear relaxation modulus is simply sketched and shown in Fig. 6.4 for a reference temperature of −12°C, which will be used to determine the Poisson's ration at this temperature.

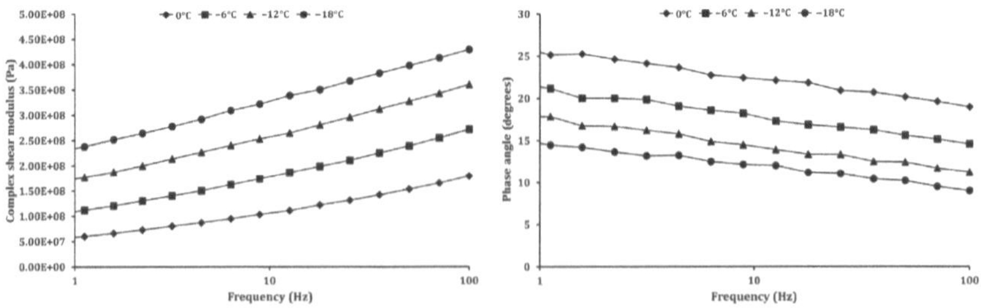

Figure 6.3 Complex shear modulus (G^*) and phase angle (δ) results of a dynamic shear rheometer test for an original asphalt binder at different test temperatures.

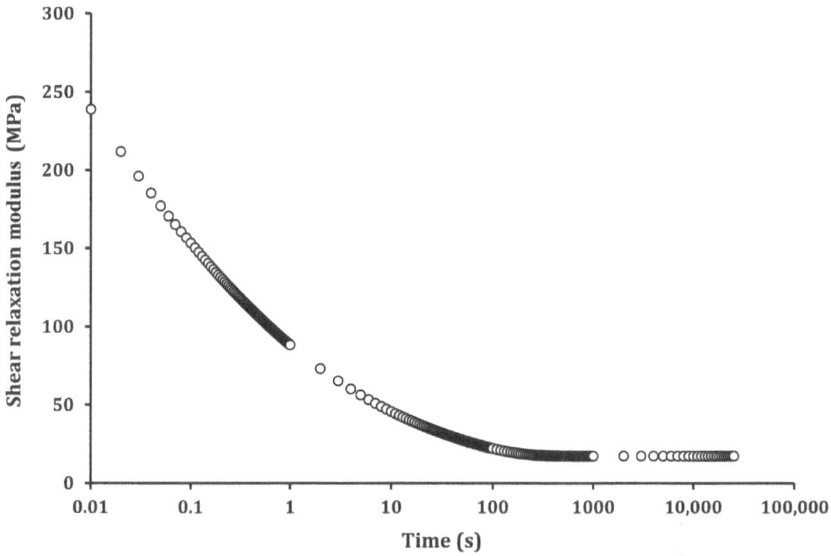

Figure 6.4 Master curve of shear relaxation modulus for the original asphalt binder at a reference temperature of −12°C.

These two parameters (G^* and δ) are used to determine the storage (G') and loss (G'') moduli of a binder as follows (see also Section 2.4):

$$G'(\omega) = |G^*|\cos\delta \tag{6.23}$$

$$G''(\omega) = |G^*|\sin\delta \tag{6.24}$$

where ω is the angular frequency (1/s). With regard to Eq. (2.24), it is possible to define the storage modulus in terms of the relaxation modulus in the frequency domain as:

$$G'(\omega) = \int_0^{\infty} \omega\ G(\eta)\sin(\omega\eta)d\eta \tag{6.25}$$

in which η is the variable of integration. Considering a generalized Maxwell model as introduced in Section 3.2.4 and substituting its relaxation modulus function in the form of a Prony series [Eq. (3.25)] in Eq. (6.25), this equation can be rewritten as:

$$G'(\omega) = \int_0^{\infty} \omega \sin(\omega\eta) \underbrace{\left[\sum_{i=1}^{n} G_i \exp\left(-\frac{\eta}{\tau_i}\right)\right]}_{G(\eta)} d\eta = \sum_{i=1}^{n} G_i\omega \int_0^{\infty} \sin(\omega\eta)\exp\left(-\frac{\eta}{\tau_i}\right)d\eta \tag{6.26}$$

The Laplace transforms of sin(t) and sin(ωt) are:

$$L\{\sin(t)\} = \int_0^\infty e^{-st}\sin(t)dt = \frac{1}{s^2+1}$$
$$L\{\sin(\omega t)\} = \int_0^\infty e^{-(s\omega)t}\sin(\omega t)dt = \frac{\omega}{s^2+\omega^2} \tag{6.27}$$

Thus the integral on the right-hand side of Eq. (6.26) can be considered as a Laplace transform based on Eq. (6.27) and $G'(\omega)$ is derived as follows:

$$G'(\omega) = \sum_{i=1}^{n} G_i\omega\frac{\omega}{\frac{1}{\tau_i^2}+\omega^2} = \sum_{i=1}^{n}\frac{G_i\omega^2\tau_i^2}{1+\omega^2\tau_i^2} \tag{6.28}$$

Similar to the relation for the storage modulus, the loss modulus can be expressed in terms of the relaxation modulus in the frequency domain as:

$$G''(\omega) = \int_0^\infty \omega\, G(\eta)\cos(\omega\eta)d\eta \tag{6.29}$$

Again, by considering a generalized Maxwell model and using the equation of its relaxation modulus in the form of a Prony series, it is possible to rewrite Eq. (6.29) as:

$$G''(\omega) = \int_0^\infty \omega\cos(\omega\eta)\underbrace{\left[\sum_{i=1}^{n} G_i\exp\left(-\frac{\eta}{\tau_i}\right)\right]}_{G(\eta)}d\eta = \sum_{i=1}^{n} G_i\omega\int_0^\infty \cos(\omega\eta)\exp\left(-\frac{\eta}{\tau_i}\right)d\eta \tag{6.30}$$

The Laplace transforms of cos(t) and cos(ωt) are:

$$L\{\cos(t)\} = \int_0^\infty e^{-st}\cos(t)dt = \frac{s}{s^2+1}$$
$$L\{\cos(\omega t)\} = \int_0^\infty e^{-(s\omega)t}\cos(\omega t)dt = \frac{s}{s^2+\omega^2} \tag{6.31}$$

Then, the integral on the right-hand side of Eq. (6.30) can be considered as a Laplace transform based on Eq. (6.31) and $G''(\omega)$ would be:

$$G''(\omega) = \sum_{i=1}^{n} G_i\omega\frac{1/\tau_i}{\frac{1}{\tau_i^2}+\omega^2} = \sum_{i=1}^{n} G_i\frac{\tau_i\omega}{1+\omega^2\tau_i^2} \tag{6.32}$$

Having determined these parameters (G_i as spring constants and τ_i is relaxation time in the generalized Maxwell model) by fitting both Eqs. (6.28) and (6.32) on experimental data of $G'(\omega)$ and $G''(\omega)$ simultaneously, the relaxation shear modulus can be simply written in the time domain as:

$$G(t) = G_\infty + \sum_{i=1}^{n} G_i \exp\left(-t/\tau_i\right) \tag{6.33}$$

As mentioned in Section 4.3.1, using the BBR test, the flexural creep stiffness, $S(t)$, of a binder can be determined as:

$$S(t) = \frac{PL^3}{4bh^3\delta(t)} \tag{6.34}$$

where $\delta(t)$ is the beam deflection at mid-span (mm), P is the constant concentrated load at the beam mid-span (N), L is the length of beam span (mm), b and h are the width and height of the beam (mm), respectively, and t is time (seconds). Fig. 6.5 shows creep compliance $[D(t) = 1/S(t)]$ at different test temperatures of 0°C, −6°C, −12°C, and −18°C for the original asphalt binder for which the complex shear modulus and phase angle are shown in Fig. 6.3. Similar to the master curve shown for the shear relaxation modulus, the master curve of the relaxation modulus is

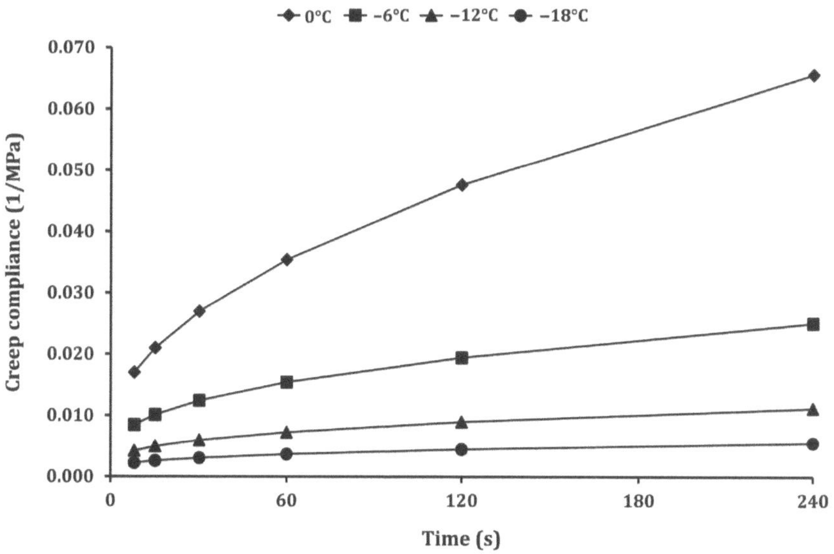

Figure 6.5 Creep compliance results of a bending beam rheometer for the original asphalt binder at different test temperatures.

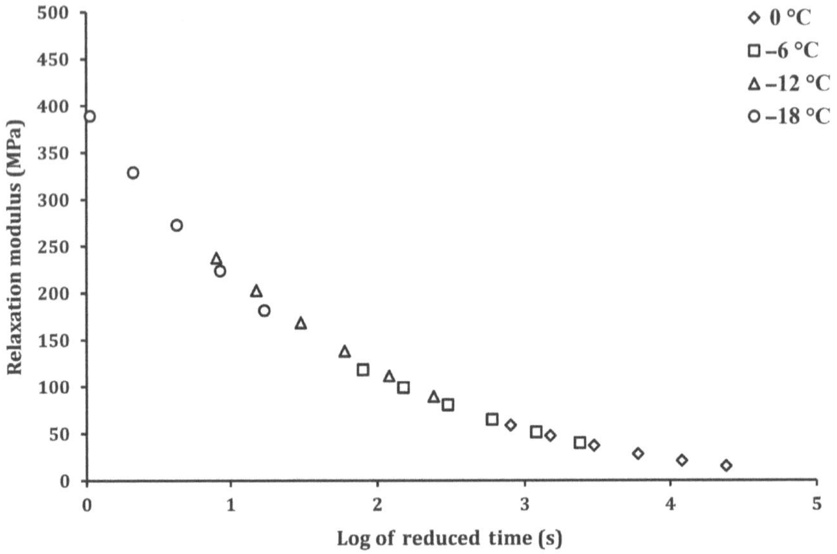

Figure 6.6 Master curve of relaxation modulus for the original asphalt binder at a reference temperature of −12°C.

depicted in Fig. 6.6 for the original asphalt binder at a reference temperature of −12°C.

Using a simple fractional viscoelastic model (see Section 3.3.2) for explaining creep compliance, $D(t) = At^{\alpha}$ (Hajikarimi et al., 2013), it is possible to determine the relaxation modulus, $E(t)$ as:

$$E(t) = \frac{t^{-\alpha}}{A\,\Gamma(1+\alpha)\,\Gamma(1-\alpha)} \tag{6.35}$$

where A and α are constant parameters determined by fitting $D(t) = At^{\alpha}$ on the curve of creep compliance versus time and $\Gamma(.)$ is a Gamma function. This procedure will be described comprehensively in Section 6.4 as a method of interconversion. Given Eqs. (6.33) and (6.35), a DSR frequency sweep test and a BBR three-point bending test can generate the data necessary to determine $G(t)$ and $E(t)$, respectively. In an elastic medium, the relation between G and E is as presented in Eq. (6.22). Applying the CP to this equation, it is possible to derive Eq. (6.36) as:

$$\overline{G}(s) = \frac{\overline{E}(s)}{2(1 + s\overline{\nu}(s))} \tag{6.36}$$

in which $\bar{G}(s)$, $\bar{E}(s)$, and $\bar{\nu}(s)$ are the inversed form of relaxation shear modulus, relaxation modulus, and Poisson's ratio in the Laplace domain, respectively. Rearranging and simplifying Eq. (6.36) yields:

$$\overline{E}(s) = 2(1 + s\overline{\nu}(s))\overline{G}(s) = 2\overline{G}(s) + 2\left(s\overline{\nu}(s) - \nu_g\right)\overline{G}(s) + 2\nu_g\overline{G}(s) \quad (6.37)$$

which can then be rewritten as:

$$E(t) = 2\left(1 + \nu_g\right)G(t) + 2\int_0^t G(t-\xi)\frac{d\nu}{d\xi}d\xi \quad (6.38)$$

where ν_g is the initial value of the Poisson's ratio and ξ is an auxiliary variable of integration. Considering a time domain like $[0, t]$, assumed as $[t_0, t_N]$, Eq. (6.38) can be rewritten in a discrete form using a time step as:

$$E(t_n) = 2\left(1 + \nu_g\right)G(t_n) + 2\sum_{i=1}^{n}\int_{t_{i-1}}^{t_i} G(t_n-\xi)\frac{d\nu}{d\xi}d\xi \quad (6.39)$$

Implementing the trapezoidal rule, the integration of Eq. (6.39) can then be determined as:

$$E(t_n) = 2\left(1 + \nu_g\right)G(t_n) + 2\sum_{i=1}^{n}\frac{G(t_n - t_i) + G(t_n - t_{i-1})}{2}[\nu(t_i) - \nu(t_{i-1})] \quad (6.40)$$

For calculating the Poisson's ratio based on $G(t)$ and $E(t)$, considering $j = i-1$, Eq. (6.40) is rewritten as:

$$\begin{aligned}
\frac{E(t_n) - 2\left(1 + \nu_g\right)G(t_n)}{2} &= \sum_{i=1}^{n}\frac{G(t_n - t_i) + G(t_n - t_{i-1})}{2}[\nu(t_i) - \nu(t_{i-1})] \\
&= \sum_{i=1}^{n}\frac{G(t_n - t_i) + G(t_n - t_{i-1})}{2}\nu(t_i) \\
&\quad - \sum_{j=0}^{n-1}\frac{G\left(t_n - t_{j+1}\right) + G\left(t_n - t_j\right)}{2}\nu\left(t_j\right) \\
&= \frac{G(0) + G(t_n - t_{n-1})}{2}\nu(t_n) \\
&\quad - \frac{G(t_n - t_1) + G(t_n - t_0)}{2}\nu_g \\
&\quad + \sum_{i=1}^{n-1}\frac{G(t_n - t_{i-1}) - G(t_n - t_{i+1})}{2}\nu(t_i)
\end{aligned} \quad (6.41)$$

and the Poisson's ratio can be calculated by reordering Eq. (6.41) as:

$$\nu(t_n) = \frac{2}{G(0) + G(t_n - t_{n-1})} \left[\begin{array}{c} \frac{E(t_n) - 2(1+\nu_g)G(t_n)}{2} + \frac{G(t_n - t_1) + G(t_n - t_0)}{2}\nu_g \\ - \sum_{i=1}^{n-1} \frac{G(t_n - t_{i-1}) - G(t_n - t_{i+1})}{2}\nu(t_i) \end{array} \right] \tag{6.42}$$

where the variables are as previously defined and:

$$\nu_g = \frac{E(0)}{2G(0)} - 1 \tag{6.43}$$

To calculate $\nu(t)$ for the original asphalt [$G(t)$ and $E(t)$ of this original asphalt binder are shown in Figs 6.4 and 6.6 at a reference temperature of −12°C] based on Eq. (6.42) a MATLAB code was developed that can set different time steps to automatically calculate $\nu(t)$ in a recursive manner. To investigate the effect of time step size on the calculated $\nu(t)$, four different time step sizes were considered: 0.1, 1, 5, and 10 s for calculating $\nu(t)$ at $t = 100$ s. It is evident from the figure that increasing the time step size can significantly affect the $\nu(t)$ results. Therefore it is important to select a time step size small enough to converge to a specific value. It is apparent that using inappropriate time step sizes can lead to poor $\nu(t)$.

To facilitate using the method for determining Poisson's ratio presented in this section, the MATLAB code used is represented in Box 6.2. Two functions of "BBR" and "Prony" are used in this code, which are shown in Boxes 6.3 and 6.4 for calculating relaxation modulus, $E(t)$, and shear relaxation modulus, $G(t)$, based on results of BBR and DSR test, respectively.

6.3 System of linear algebraic equations based on Prony series

The following relationship between the relaxation modulus, $E(t)$, and the creep compliance, $D(t)$, for linear viscoelastic materials can be obtained as an integral expression (Park & Kim, 1999):

$$\int_0^t E(t-\xi)\frac{dD(\xi)}{d\xi}d\xi = 1 \tag{6.44}$$

In this section, a Prony series is used to interconvert relaxation modulus and creep compliance based on Eq. (6.44). Therefore the generalized Maxwell or Kelvin–Voigt introduced in Section 3.2.4 are used. The

BOX 6.2 MATLAB code for determination of the Poisson's ration having results of DSR and BBR test at low temperatures by using Hopkins and Hamming's method.

```
%Finding the Poisson's ratio based on G(t) and E(t)

t_n = input('desired time for determining Poisson's ratio = ');
m = input('time step = ');
t = 0:m:t_n;
t = t';
v = [0.0;0.0];
vg = input ('initial Poisson's ratio = ');
t0 = 0;
for i = 2:t_n/m
    v_T = 0;
    for j = 1:i-1
        v_T   = v_T + ((Prony(i*m-t(j),E_i,to_i)-Prony(i*m-t(j + 2),
        E_i,to_i))*v(j)/2);
    end
    v_T1  =  (2/((Prony(0,E_i,to_i) + Prony(m,E_i,to_i))))*(((-1)*
    v_T) +...
        ((BBR(i*m)-2*(1 + vg)*Prony(i*m,E_i,to_i))/2) +...
        ((Prony(i*m-t(1),E_i,to_i) + Prony(i*m-t0,E_i,to_i))*vg/2));
    v(i + 1) = v_T1;
end
```

BOX 6.3 MATLAB function "Prony" for determination of shear relaxation modulus, *G*(*t*), based on Prony series for a generalized Maxwell model.

```
%G_i and to_i are two vectors which their size is equal and depended
on %the selected generalized Maxwell model

function G = Prony(t,G_i,to_i)
G = G0;
for j = 1:size(G_i,1)
    g(j) = G_i(j)*exp(-t/to_i(j));
    G = G+ g(j);
end
end
```

BOX 6.4 MATLAB function "BBR" for determination of relaxation modulus, *E*(*t*), based on a simple fractional element.

```
%A = 393, and alpha = -0.2592 for a simple fractional element

function E = BBR(t)
E = 393*t^(-.2592);
end
```

creep compliance of a linear viscoelastic material can be characterized by using the generalized Kelvin–Voigt model as follows:

$$D(t) = D_g + \frac{t}{\eta_0} + \sum_{j=1}^{n} D_j\left(1 - \exp\left(-t/\rho_j\right)\right) \tag{6.45}$$

in which, following positive parameters, $D(t)$, D_g, η_0, D_j, and ρ_j are the creep compliance, glassy compliance, zero-shear viscosity, retardation strength, and retardation times, respectively. The combination of springs and dashpots related to the linear viscoelastic model of the generalized Kelvin–Voigt is shown in Fig. 3.1. In Eq. (6.45), it is assumed that the creep compliance data is available and D_g, D_j, and ρ_j are fitted parameters using the curve-fitting toolbox of MATLAB (CFTOOL) according to the principle of multiple nonlinear regression. For solid-like viscoelastic materials, t/η_0 is relatively small and negligible. Thus Eq. (6.45) can be approximated as:

$$D(t) = D_g + \sum_{j=1}^{n} D_j\left(1 - \exp\left(-t/\rho_j\right)\right) \tag{6.46}$$

Throughout a curve-fitting procedure, parameters (D_g and n series of D_j and ρ_j where n is the number of Kelvin–Voigt components) can be determined and the relaxation modulus can be calculated based on the creep compliance data by the following equation, the so-called Prony series:

$$E(t) = E_e + \sum_{i=1}^{m} E_i \exp\left(-t/\tau_i\right) \tag{6.47}$$

in which E_e and E_i are the equilibrium modulus and the relaxation strength, respectively. To solve Eq. (6.44), both viscoelastic solids ($E_e > 0$ and $\eta_0 \rightarrow \infty$) and viscoelastic liquids ($E_e = 0$ and η_0 is finite) can be considered.

6.3.1 Viscoelastic solids

For viscoelastic solids, substituting the Prony series of Eqs. (6.46) and (6.47) into Eq. (6.44) would be:

$$\int_0^t \left(E_e + \sum_{i=1}^{m} E_i \exp\left(-\frac{(t-\xi)}{\tau_i}\right)\right)\left(D_g\delta(\xi) + \sum_{j=1}^{n} \frac{D_j}{\rho_j}\exp\left(-\frac{\xi}{\rho_j}\right)\right)d\xi = 1 \tag{6.48}$$

or,

$$\begin{aligned} &D_g\left(E_e + \sum_{i=1}^{m} E_i \exp\left(-\frac{t}{\tau_i}\right)\right) + E_e\sum_{j=1}^{n}\frac{D_j}{\rho_j}\int_0^t \exp\left(-\frac{\rho}{\rho_j}\right)d\xi \\ &+\sum_{i=1}^{m}\sum_{j=1}^{n}\frac{E_iD_j}{\rho_j}\exp\left(-\frac{t}{\tau_i}\right)\int_0^t \exp\left(-\left[\left(\frac{\xi}{\rho_j}\right)-\left(\frac{\xi}{\tau_i}\right)\right]\right)d\xi = 1 \end{aligned} \tag{6.49}$$

in which $\delta(.)$ denotes the Dirac delta function. Two integrals in Eq. (6.49) can be simply determined as:

$$\int_0^t \exp\left(-\frac{\rho}{\rho_j}\right)d\xi = \rho_j\left[1 - \exp\left(-\frac{t}{\rho_j}\right)\right] \tag{6.50}$$

$$\begin{aligned} &\int_0^t \exp\left(-\left[\left(\frac{\xi}{\rho_j}\right)-\left(\frac{\xi}{\tau_i}\right)\right]\right) \\ &d\xi = \frac{\tau_i\rho_j}{\tau_i-\rho_j}\left(1-\exp\left(-\left[\left(\frac{\xi}{\rho_j}\right)-\left(\frac{\xi}{\tau_i}\right)\right]\right)\right), \quad \tau_i \neq \rho_j \\ &\int_0^t \exp\left(-\left[\left(\frac{\xi}{\rho_j}\right)-\left(\frac{\xi}{\tau_i}\right)\right]\right)d\xi = t \quad , \quad \tau_i = \rho_j \end{aligned} \tag{6.51}$$

Substituting Eqs. (6.51) and (6.50) into Eq. (6.49) and rearranging, it would be:

$$\begin{aligned} &E_i\sum_{i=1}^{m}\left[D_g\exp\left(-\frac{t}{\tau_i}\right)\right. \\ &\left.+\sum_{j=1}^{n}D_j\exp\left(-\frac{t}{\tau_i}\right)\frac{\tau_i}{\tau_i-\rho_j}\left(1-\exp\left(-\left[\left(\frac{t}{\rho_j}\right)-\left(\frac{t}{\tau_i}\right)\right]\right)\right)\right] \\ &= 1 - E_e\left(D_g + \sum_{j=1}^{n}D_j\left[1-\exp\left(-\frac{t}{\rho_j}\right)\right]\right) \end{aligned} \tag{6.52}$$

for $\tau_i \neq \rho_j$. It can be readily derived that:

$$E_i \sum_{i=1}^{m}\left[D_g \exp\left(-\frac{t}{\tau_i}\right)+\sum_{j=1}^{n}\frac{D_j t}{\rho_j}\exp\left(-\frac{t}{\tau_i}\right)\right] = 1 - E_e\left(D_g+\sum_{j=1}^{n}D_j\left[1-\exp\left(-\frac{t}{\rho_j}\right)\right]\right) \tag{6.53}$$

for $\tau_i = \rho_j$.

When $D(t)$ is known and $E(t)$ is the desired target function (determination of the relaxation modulus based on creep compliance data), the initial-value and final-value problems of the Laplace transform can be used to determine E_e as (Churchill, 1972):

$$E_e \cong \lim_{t\to 0} E(t) = \lim_{s\to\infty} \overline{E}(s) = \frac{1}{\lim_{s\to\infty}\overline{D}(s)} = \frac{1}{\lim_{t\to 0} D(t)} = \frac{1}{D_g + \sum_{j=1}^{n} D_j} \tag{6.54}$$

For the case when $E(t)$ is known and $D(t)$ is the unknown target function, it is possible to rearrange Eq. (6.49) by substituting Eqs. (6.50) and (6.51) as:

$$D_j \sum_{j=1}^{n}\left[E_e\left(1-\exp\left(-\frac{t}{\tau_i}\right)\right)+\sum_{i=1}^{m}E_i\frac{\tau_i}{\tau_i-\rho_j}\left(\exp\left[\left(\frac{t}{\tau_i}\right)-\left(\frac{t}{\rho_j}\right)\right]\right)\right] = 1 - D_g\left(E_e+\sum_{i=1}^{m}E_i \exp\left(-\frac{t}{\tau_i}\right)\right) \tag{6.55}$$

for $\tau_i \neq \rho_j$. It can be simply derived that:

$$D_j \sum_{j=1}^{n}\left[E_e\left(1-\exp\left(-\frac{t}{\tau_i}\right)\right)+\sum_{i=1}^{m}\frac{E_i t}{\rho_j}\exp\left(-\frac{t}{\tau_i}\right)\right] = 1 - D_g\left(E_e+\sum_{i=1}^{m}E_i \exp\left(-\frac{t}{\tau_i}\right)\right) \tag{6.56}$$

for $\tau_i = \rho_j$. Similar to the procedure which is used to calculate E_e, it is possible to determine D_g as follows:

$$D_g \cong \lim_{t\to 0} D(t) = \lim_{s\to\infty} \overline{D}(s) = \frac{1}{\lim_{s\to\infty}\overline{E}(s)} = \frac{1}{\lim_{t\to 0} E(t)} = \frac{1}{E_e + \sum_{i=1}^{m} E_i} \tag{6.57}$$

Having one set of constants, either $\{E_e, E_i\ (i=1, 2, 3, \ldots, m), \text{and } \tau_i\}$ for the relaxation modulus or $\{D_g, D_j\ (j=1, 2, 3, \ldots, n), \tau_i, \text{and } \eta_0\}$ for the creep compliance as well as target time constants, the other unknown set of constants can be readily determined by solving the subsequent system of linear algebraic equations. When $D(t)$ is known, the desired relaxation modulus can be determined according to the following system of equations:

$$[A]\{E\} = \{B\} \tag{6.58}$$

in which:

$$A_{ki} = D_g \exp\left(-\frac{t_k}{\tau_i}\right) + \sum_{j=1}^{n} \frac{D_j \tau_i}{\tau_i - \rho_j}\left[\exp\left(-\frac{t_k}{\tau_i}\right) - \exp\left(-\frac{t_k}{\rho_i}\right)\right],\quad \tau_i \neq \rho_j$$

or

$$A_{ki} = D_g \exp\left(-\frac{t_k}{\tau_i}\right) + \sum_{j=1}^{n} \frac{D_j t_k}{\rho_j},\quad \tau_i = \rho_j \tag{6.59}$$

and,

$$B_k = 1 - \frac{D_g + \sum_{j=1}^{n} D_j\left[1 - \exp\left(-t_k/\rho_j\right)\right]}{D_g + \sum_{j=1}^{n} D_j} \tag{6.60}$$

In Eqs. (6.59) and (6.60), t_k denotes a discrete time corresponding to the time represented in Eq. (6.44). On the other hand, when $E(t)$ is known, the desired creep compliance can be determined according to the following system of equations:

$$[A]\{D\} = \{B\} \tag{6.61}$$

in which:

$$A_{kj} = E_e\left(1 - \exp\left(-\frac{t_k}{\rho_j}\right)\right) + \sum_{i=1}^{m} \frac{E_i \tau_i}{\tau_i - \rho_j}\left[\exp\left(-\frac{t_k}{\tau_i}\right) - \exp\left(-\frac{t_k}{\rho_j}\right)\right],\quad \tau_i \neq \rho_j$$

or

$$A_{kj} = E_e\left(1 - \exp\left(-\frac{t_k}{\rho_j}\right)\right) + \sum_{i=1}^{m} \frac{E_i t_k}{\rho_j},\quad \tau_i = \rho_j \tag{6.62}$$

and

$$B_k = 1 - \frac{E_e + \sum_{i=1}^{m} E_i \exp\left(-t_k/\tau_j\right)}{E_e + \sum_{i=1}^{m} E_i} \tag{6.63}$$

where t_k again denotes a discrete time corresponding to the time represented in Eq. (6.44).

6.3.2 Viscoelastic liquids

For viscoelastic liquids, substituting the Prony series of Eqs. (6.46) and (6.47) into Eq. (6.44) yields the following equation:

$$\int_0^t \left(\sum_{i=1}^{m} E_i \exp\left(-\frac{(t-\xi)}{\tau_i}\right)\right)\left(D_g\delta(\xi) + \frac{1}{\eta_0} + \sum_{j=1}^{n}\frac{D_j}{\rho_j}\exp\left(-\frac{\xi}{\rho_j}\right)\right)d\xi = 1 \quad (6.64)$$

or,

$$\begin{aligned} &D_g\sum_{i=1}^{m} E_i \exp\left(-\frac{t}{\tau_i}\right) + \frac{1}{\eta_0}\sum_{i=1}^{m} E_i \int_0^t \exp\left(-\frac{(t-\xi)}{\tau_i}\right)d\xi \\ &+ \sum_{i=1}^{m}\sum_{j=1}^{n}\frac{E_iD_j}{\rho_j}\exp\left(-\frac{t}{\tau_i}\right)\int_0^t \exp\left(-\left[\left(\frac{\xi}{\rho_j}\right)-\left(\frac{\xi}{\tau_i}\right)\right]\right)d\xi = 1 \end{aligned} \quad (6.65)$$

in which $\delta(.)$ denotes the Dirac delta function. Two integrals in Eq. (6.65) can be simply determined as:

$$\int_0^t \exp\left(-\frac{(t-\xi)}{\tau_i}\right)d\xi = \tau_i\left[1 - \exp\left(-\frac{t}{\tau_i}\right)\right] \quad (6.66)$$

$$\begin{aligned} &\int_0^t \exp\left(-\left[\left(\frac{\xi}{\rho_j}\right)-\left(\frac{\xi}{\tau_i}\right)\right]\right)d\xi = \frac{\tau_i\rho_j}{\tau_i-\rho_j}\left(1-\exp\left(-\left[\left(\frac{\xi}{\rho_j}\right)-\left(\frac{\xi}{\tau_i}\right)\right]\right)\right), \quad \tau_i \neq \rho_j \\ &\int_0^t \exp\left(-\left[\left(\frac{\xi}{\rho_j}\right)-\left(\frac{\xi}{\tau_i}\right)\right]\right)d\xi = t, \quad \tau_i = \rho_j \end{aligned} \quad (6.67)$$

Then, by substituting Eqs. (6.66) and (6.67) into Eq. (6.65) and rearranging, the following equation can be simply written:

$$\begin{aligned} &D_j\sum_{j=1}^{n}\left[\sum_{i=1}^{m}\frac{E_i\tau_i}{\tau_i-\rho_j}\left(\exp\left(-\frac{t}{\tau_i}\right)-\exp\left(-\frac{t}{\rho_j}\right)\right)\right] = 1 - D_g\sum_{i=1}^{m}E_i \\ &\exp\left(-\frac{t}{\tau_i}\right) - \frac{1}{\eta_0}\sum_{i=1}^{m}E_i\tau_i\left(1-\exp\left(-\frac{t}{\tau_i}\right)\right) \end{aligned} \quad (6.68)$$

for $\tau_i \neq \rho_j$. It can be simply derived that:

$$\begin{aligned} &D_j\sum_{j=1}^{n}\left[\sum_{i=1}^{m}\frac{E_it}{\rho_j}\exp\left(-\frac{t}{\tau_i}\right)\right] = 1 - D_g\sum_{i=1}^{m}E_i \\ &\exp\left(-\frac{t}{\tau_i}\right) - \frac{1}{\eta_0}\sum_{i=1}^{m}E_i\tau_i\left(1-\exp\left(-\frac{t}{\tau_i}\right)\right) \end{aligned} \quad (6.69)$$

for $\tau_i = \rho_j$. Eq. (6.57) can be reduced for viscoelastic liquids considering $E_e = 0$ as follows:

$$D_g \cong \frac{1}{\sum_{i=1}^{m} E_i} \tag{6.70}$$

The constant η_0 can also be written in the form of E_i constants as follows:

$$\eta_0 = \sum_{i=1}^{m} \tau_i E_i \tag{6.71}$$

6.3.3 Numerical example

In this section, the results of a BBR test is used to show the ability of the presented method to determine the relaxation modulus of a modified asphalt binder based on its creep compliance data. According to ASTM D6648, by using the BBR test, mechanical and rheological properties of neat and modified asphalt binders can be evaluated by determining the flexural creep stiffness of samples at different low test temperatures. Deflection data at the mid-span is acquired in 0.5 s time steps for 240 s. Using Eq. (6.34), the flexural creep stiffness is determined for an asphalt binder modified by introducing 3% of styrene–butadiene–styrene (SBS). Then, the creep compliance is calculated by using the following equation:

$$D(t) = \frac{1}{S(t)} \tag{6.72}$$

By fitting Eq. (6.45) for a generalized Kelvin–Voigt model (having nine Kelvin–Voigt components) on experimental creep compliance data [$D(t)$ vs time], the glassy compliance, D_g, retardation strengths, D_j, and retardation times, ρ_j, can be determined as shown in Table 6.1. It is necessary to mention that the abovementioned fitting is done on the master curve of creep compliance at a reference temperature of −18°C to widen the time range. According to the selected times and known parameters, matrices A and B can be derived based on Eqs. (6.59) and (6.60), respectively. Hence, nine parameters of the relaxation strengths can be determined by solving Eq. (6.58). The resulting parameters are tabulated in Table 6.1 including E_e, E_i, and τ_i. Consequently, the relaxation modulus corresponding to the creep compliance data for this modified asphalt binder can be determined, which is indicated in Fig. 6.7. It can be seen in this figure that there is a significant difference between flexural creep stiffness, $S(t)$, and the calculated relaxation modulus, $E(t)$ which proves

Table 6.1 Fitted and calculated parameters for styrene–butadiene–styrene–modified asphalt binder by using a system of algebraic equations based on Prony series.

D_i Retardation strength (1/MPa)		D_g Glassy compliance (1/MPa)	ρ_j Retardation time (s)		E_e Equilibrium modulus (MPa)	τ_i Relaxation time (s)		t_k Selected time (s)	E_i Relaxation strength (MPa)	
D1	6.00E − 04	5.75E − 04	τ1	2.54E + 01	1.12E + 02	ρ1	8.91E − 04	0.5	E1	4.87E + 08
D2	2.86E − 04		τ2	9.47E − 02		ρ2	8.91E − 03	2	E2	−5.12E + 07
D3	3.86E − 04		τ3	7.12E − 01		ρ3	8.91E − 02	4	E3	2.50E + 05
D4	3.28E − 04		τ4	2.07E + 00		ρ4	8.91E − 01	8	E4	4.11E + 02
D5	4.75E − 04		τ5	3.52E − 01		ρ5	8.91E + 00	15	E5	1.14E + 02
D6	5.95E − 04		τ6	1.06E + 01		ρ6	8.91E + 01	30	E6	2.37E + 02
D7	2.10E − 04		τ7	2.33E − 06		ρ7	8.91E + 02	60	E7	−2.96E + 03
D8	2.48E − 05		τ8	2.69E + 01		ρ8	8.91E + 03	120	E8	2.63E + 04
D9	5.41E − 03		τ9	2.77E + 02		ρ9	8.91E + 04	240	E9	−2.34E + 04

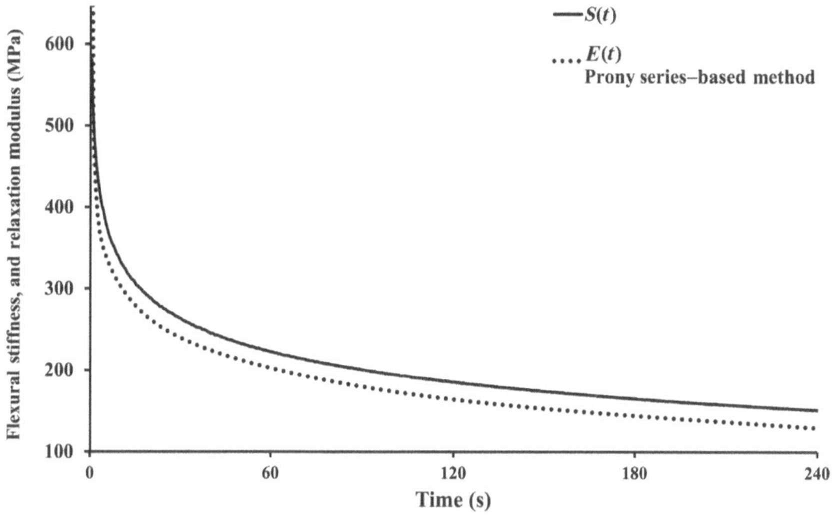

Figure 6.7 The flexural creep stiffness and the relaxation modulus curve for styrene–butadiene–styrene–modified asphalt binder at a reference temperature of −18°C determined by system of algebraic equations based on Prony series.

that in the time domain: $E(t) \neq 1/D(t)$. Therefore if the Prony series is required to be used in a numerical framework (e.g., for simulating viscoelastic materials in ABAQUS), it is necessary to determine the relaxation modulus accurately and it is not possible to approximate it considering a straightforward relation between creep compliance and relaxation modulus in the time domain.

The MATLAB code which is developed for fitting the creep compliance function on experimental data and then solving the system of the algebraic equation to determine the relaxation modulus is presented here in Box 6.5. In this code, it is evident that both t and S are matrices containing experimental results of the BBR test as the main input for calculating the relaxation modulus.

6.4 General approximation methods of interconversion

Two approximation methods of Laplace transform inversions were presented by Schapery (1962): the direct method and the collocation method. The Carson transform (s-multiplied Laplace transform) of the uniaxial relaxation modulus can be defined as:

$$\tilde{E}(s) = s\int_0^\infty E(t)\exp[-st]dt = s\int_0^\infty E(t)t\exp[-st]d(\ln t) \qquad (6.73)$$

BOX 6.5 MATLAB code for determination of the relaxation modulus having the creep compliance data by using a system of algebraic equations based on the Prony series.

```
%Using system of algebraic equations to determine the relaxation %
modulus based on the creep compliance data for a particular case %
where 9 components are considered for the generalized Kelvin-Voigt
%model

clc
clear f i p1 p2 Di Einf s j Dbar Ebar delta Ei tol tolmax Tau g Ro E D0 D
%Import Time As t
%Import Stiffness As S
Ro = [0.001 0.01 0.1 1 10 100 1000 10000 100000];
D = 1./S;
while 1 > 0
    [f,p1] = fit(t,D,'D0 + D1*(1-exp(-x/0.001)) + D2*(1-exp(-x/
    0.01)) + D3*(1-exp(-x/0.1)) + D4*(1-exp(-x/1)) + D5*(1-exp(-x/
    10)) + D6*(1-exp(-x/100)) + D7*(1-exp(-x/1000)) + D8*(1-exp(-x/
    10000)) + D9*(1-exp(-x/100000))','lower',
    [0,0,0,0,0,0,0,0,0]);
    Di = coeffvalues(f);
    if Di(1) > 0 && Di(2) > 0 && Di(3) > 0 && Di(4) > 0 && Di(5) > 0 && Di
    (6) > 0 && Di(7) > 0 && Di(8) > 0 && Di(9) > 0 && Di(10) > 0
        break
    end
end
D0 = Di(1);
Di = [Di(2) Di(3) Di(4) Di(5) Di(6) Di(7) Di(8) Di(9) Di(10)];
Einf = 1/(D0 + sum(Di))
s = [];
Dbar = [];
Ebar = [];
Tau = [0.001 0.01 0.1 1 10 100 1000 10000 100000];
E = [];
for j = 1:200
    s = [s transpose(linspace(2*j-1,2*j + 1,401))];

Dbar = [Dbar,D0 + Di(1)./(s(:,j)*0.001 + 1) + Di(2)./(s(:,j)*
0.01 + 1) + Di(3)./(s(:,j)*0.1 + 1) + Di(4)./(s(:,j)*1 + 1) + Di(5)./
(s(:,j)*10 + 1) + Di(6)./(s(:,j)*100 + 1) + Di(7)./(s(:,j)*
1000 + 1) + Di(8)./(s(:,j)*10000 + 1) + Di(9)./(s(:,j)*100000 + 1)];
    Ebar = [Ebar 1./Dbar(:,j)];
    delta = Ebar-Einf;
```

(Continued)

BOX 6.5 (Continued)

```
[g,p2]=fit(s(:,j),delta(:,j),'E1*x./(x+1/0.001)+E2*x./(x+1/
0.01)+E3*x./(x+1/0.1)+E4*x./(x+1/1)+E5*x./(x+1/10)+E6*x./
(x+1/100)+E7*x./(x+1/1000)+E8*x./(x+1/10000)+E9*x./(x+1/
100000)');
    Ei=coeffvalues(g);
    E=[E,Einf+Ei(1)*exp(-t./Tau(1))+Ei(2)*exp(-t./ Tau(2))+Ei
    (3)*exp(-t./ Tau(3))+Ei(4)*exp(-t./ Tau(4))+Ei(5)*exp(-t./
    Tau(5))+Ei(6)*exp(-t./ Tau(6))+Ei(7)*exp(-t./ Tau(7))+Ei
    (8)*exp(-t./ Tau(8))+Ei(9)*exp(-t./ Tau(9))];
    if j>1
        tol=[abs(E(:,j)-E(:,j-1))];
        tolmax=max(tol);
        if tolmax<0.1
            break
        end
    end
end
```

where $\tilde{E}(s) = s\bar{E}(s)$ and $\bar{E}(s)$ is the Laplace transform of the relaxation modulus. The Carson transform has approximate interconversion properties as follows:

$$\tilde{E}(s) \cong E(t)|_{t=\frac{\alpha}{s}} \tag{6.74}$$

or

$$E(t) \cong \tilde{E}(s)|_{s=\frac{\alpha}{t}} \tag{6.75}$$

where $\alpha = \exp(-C)$ for $C = 0.5772$ (Euler's constant) resulted in $\alpha \approx 0.56$. Schapery (1962) also introduced an upgraded equation that can be used when the derivative of $\log E(t)$ with respect to $\log(t)$ is a slowly changing function of $\log(t)$ as:

$$\tilde{E}(s) \cong E(t)|_{t=\frac{\beta}{s}} \tag{6.76}$$

or

$$E(t) \cong \tilde{E}(s)|_{s=\frac{\beta}{t}} \tag{6.77}$$

in which $\beta = \{\Gamma(1-n)\}^{-(1/n)}$, $\Gamma(.)$ denotes Gamma function, and n is defined as the local log–log slope as:

$$n \equiv -\frac{d\log E}{d\log t} \equiv \frac{d\log \tilde{E}}{d\log s} \tag{6.78}$$

It is evident that when having a relaxation modulus function in the form of $E(t) \sim t^{-n}$, and $t > 0$, Eqs. (6.76) and (6.77) are exact. It is also possible to derive analogous equations for the creep compliance function by substituting the relaxation modulus, $E(t)$, in the abovementioned equations with the creep compliance, $D(t)$.

Based on Kramers–Kronig equations, Booij and Thoone (1982) proposed the following relations for converting $E''(\omega)$ to $E'(\omega)$:

$$E'(\omega) \cong E_e - \frac{\pi\omega}{2}\frac{d\left[\frac{E''(\omega)}{\omega}\right]}{d\ln\omega} \tag{6.79}$$

or,

$$E'(\omega) \cong E_e + \frac{\pi}{2}\left(1 - \frac{d\ln E''(\omega)}{d\ln\omega}\right)E''(\omega) \tag{6.80}$$

in which E_e is the equilibrium modulus. It is possible to derive these equations for the creep compliance function by replacing E' and E'' with D' and D'', respectively.

Christensen (2012) also derived an approximation between the storage modulus and the relaxation modulus as:

$$E'(\omega) \cong E(t)|_{t=\frac{2}{\pi\omega}} \tag{6.81}$$

or,

$$E(t) \cong E'(\omega)|_{\omega=\frac{2}{\pi t}} \tag{6.82}$$

Similar to other approximation relations, it is possible to replace $E'(\omega)$ by $D'(\omega)$ to develop such an equation for the creep compliance function.

6.5 Approximation method of interconversion based on general power law

According to the viscoelasticity theorem, the following exact equations can be obtained between two constitutive material functions (Tschoegl, 2012):

$$E(t) = \frac{2}{\pi}\int_0^{\infty} E'(\omega)\frac{\sin(\omega t)}{\omega}\,d\omega \tag{6.83}$$

$$E(t) = E_e + \frac{2}{\pi}\int_0^{\infty} E''(\omega)\frac{\cos(\omega t)}{\omega}d\omega \tag{6.84}$$

$$E'(\omega) = E_e + \frac{2\omega^2}{\pi}\int_0^{\infty} E''(\zeta)\frac{1}{\zeta\left(\omega^2 - \zeta^2\right)}d\zeta \tag{6.85}$$

$$E''(\omega) = \frac{2\omega}{\pi}\int_0^{\infty} [E'(\zeta) - E_e]\frac{1}{\zeta^2 - \omega^2}d\zeta \tag{6.86}$$

Eqs. (6.75) and (6.76) are known as Kronig–Kramers equations. With regard to the definitions of Carson transform represented in Eq. (6.73), it is possible to determine the Carson transform of Eqs. (6.83) and (6.84) as:

$$\tilde{E}(s) = \frac{2s}{\pi}\int_0^{\infty} E'(\omega)\frac{1}{s^2 + \omega^2}d\omega \tag{6.87}$$

$$\tilde{E}(s) = E_e + \frac{2s^2}{\pi}\int_0^{\infty} E''(\omega)\frac{1}{\omega(s^2 + \omega^2)}d\omega \tag{6.88}$$

To proceed, the materials with power law function is considered where $E(t) \sim t^{-n}$ or $E(\omega) \sim \omega^{-n}$ and $-1 < n < 1$ is a slowly varying function of $\log t$ or $\log \omega$. It is possible to write the relaxation modulus function in a general form as follows:

$$E(t) = E_n\left(\frac{t}{t_n}\right)^{-n}[1 + F(t)] \tag{6.89}$$

in which $E(t) \approx E(t_n)$ and $F = 0$ if we have a viscoelastic material with a pure power law relaxation modulus function. Actually, $F(t)$ can be an arbitrary function to specify how relaxation modulus behavior is close to the power law. Also, t_n is the time at which $-n$ [see Eq. (6.78)] is determined. The Carson transform of Eq. (6.89) can be written as:

$$\tilde{E}(s) = E_n\Gamma(1 - n)(st_n)^n[1 + c_n] \tag{6.90}$$

where

$$c_n = s^{1-n}\int_{-\infty}^{+\infty}\frac{Ft^{1-n}\exp(-st)d(\ln t)}{\Gamma(1 - n)} \tag{6.91}$$

is a relative correction factor regarding to the presence of $F(t)$ in Eq. (6.89). By neglecting c_n in Eq. (6.90) and choosing $t_n = 1/s$, Eq. (6.90) is rewritten as:

$$\tilde{E}(s) = E(t)\Gamma(1 - n) \tag{6.92}$$

Selecting a t_n which minimizes the correction factor of c_n can enhance the accuracy of the approximation. This selection depends on the value of n at the initially identified time t_n.

The storage modulus can be represented in a similar form of Eq. (6.89) as:

$$E'(\omega) = E'_n\left(\frac{\omega}{\omega_n}\right)^n\left[1 + F'\left(\frac{\omega}{\omega_n}\right)\right] \tag{6.93}$$

in which $E'_n \approx E'(\omega_n)$ and $n \approx d\log E'(\omega)/d\log(\omega)$. By neglecting F' and selecting $\omega_n = s$, the following equation can be derived:

$$\tilde{E}(s) = \frac{E'(\omega)}{\cos\left(\frac{n\pi}{2}\right)} \tag{6.94}$$

Both Eqs. (6.92) and Eq. (6.94) can be used to relate $E(t)$ and $E'(\omega)$ as:

$$E'(\omega) = \Gamma(1 - n)\cos\left(\frac{n\pi}{2}\right)E(t) \tag{6.95}$$

where $t = 1/\omega$.

By considering Eq. (6.88) and using a similar procedure, the following equations can be derived:

$$\tilde{E}(s) - E_e = \frac{E''(\omega)}{\sin\left(\frac{n\pi}{2}\right)} \tag{6.96}$$

$$E''(\omega) = \tan\left(\frac{n\pi}{2}\right)[E'(\omega) - E_e] \tag{6.97}$$

in which the local slope is:

$$n \cong \frac{d\log[E'(\omega) - E_e]}{d\log(\omega)} \tag{6.98}$$

Schapery and Park (1999) summarized the set of approximate interconversions as the following equations:

$$E(t) \cong 1\tilde{\lambda}\tilde{E}(s)|_{s=\frac{1}{t}} \quad \text{or} \quad \tilde{E}(s) \cong \tilde{\lambda}E(t)|_{t=\frac{1}{s}} \tag{6.99}$$

$$E(t) \cong \frac{1}{\lambda'}E'(\omega)|_{\omega=\frac{1}{t}} \quad \text{or} \quad E'(\omega) \cong \lambda'E(t)|_{t=\frac{1}{\omega}} \tag{6.100}$$

$$E(t) \cong \frac{1}{\lambda''}E''(\omega)|_{\omega=\frac{1}{t}} \quad \text{or} \quad E''(\omega) \cong \lambda''E(t)|_{t=\frac{1}{\omega}} \tag{6.101}$$

$$\tilde{E}(s) \cong 1\hat{\lambda}E'(\omega)|_{\omega=s} \quad \text{or} \quad E'(\omega) \cong \hat{\lambda}\tilde{E}(s)|_{s=\omega} \tag{6.102}$$

Table 6.2 Adjustment functions used in the approximate interconversion method based on the general power law.

Ratios	Adjustment function
$\tilde{\lambda} = \tilde{E}(s)/E(t)$	$\tilde{\lambda} = \Gamma(1-n)$
$\lambda' = E'(\omega)/E(t)$	$\lambda' = \Gamma(1-n)\cos\left(\frac{n\pi}{2}\right)$
$\lambda'' = E''(\omega)/E(t)$	$\lambda'' = \Gamma(1-n)\sin\left(\frac{n\pi}{2}\right)$
$\hat{\lambda} = E'(\omega)/\tilde{E}(s)$	$\hat{\lambda} = \cos\left(\frac{n\pi}{2}\right)$
$\overline{\lambda} = E''(\omega)/\tilde{E}(s)$	$\overline{\lambda} = \sin\left(\frac{n\pi}{2}\right)$
$\lambda^* = E''(\omega)/E'(\omega)$	$\lambda^* = \tan\left(\frac{n\pi}{2}\right)$

$$\tilde{E}(s) \cong 1\overline{\lambda}E''(\omega)|_{\omega=s} \quad \text{or} \quad E''(\omega) \cong \overline{\lambda}\tilde{E}(s)|_{s=\omega} \tag{6.103}$$

$$E'(\omega) \cong \frac{1}{\lambda^*}E''(\hat{\omega})|_{\hat{\omega}=\omega} \quad \text{or} \quad E''(\omega) \cong \lambda^* E'(\hat{\omega})|_{\hat{\omega}=\omega} \tag{6.104}$$

in which all types of λ variable including λ', λ'', etc. are introduced in Table 6.2 for the approximate interconversion method based on the general power law.

By applying Eq. (6.99) on both the relaxation modulus, $E(t)$, and the creep compliance, $D(t)$, and using the exact relationship between the Carson transformed function as:

$$\tilde{E}(s)\tilde{D}(s) = 1 \tag{6.105}$$

it is possible to derive the Ferry's (1980) equation by assuming a local power law behavior as follows:

$$E(t)D(t) = \Gamma(1-n)\Gamma(1+n) = \frac{\sin(n\pi)}{n\pi} \tag{6.106}$$

in which both $E(t)$ and $D(t)$ are assumed to be a power law function like the constitutive equations represented for a simple fractional element in Chapter 3, Mechanical Models of Viscoelasticity. Hajikarimi et al. (2018) demonstrated that for a simple fractional element, the relaxation modulus of asphalt binder can be determined based on results of the BBR test at low test temperatures as:

$$E(t) = S(t)\Gamma(1-n)\Gamma(1+n) \tag{6.107}$$

where $S(t)$ is flexural creep stiffness $[S(t) = 1/D(t)]$ and $D(t) = At^{-n}$. It is evident that Eq. (6.107) is in consistent with Eq. (6.106). It can be seen that by using Eq. (6.107), it is possible to determine the relaxation modulus analytically and there is no need to perform a numerical procedure which is needed for performing Hopkins and Hamming's method or the system of algebraic equations based on the Prony series.

References

Booij, H. C., & Thoone, G. P. J. M. (1982). Generalization of Kramers-Kronig transforms and some approximations of relations between viscoelastic quantities. *Rheologica Acta, 21*(1), 15–24. Available from https://doi.org/10.1007/BF01520701.

Christensen, R. (2012). *Theory of viscoelasticity: An introduction*. Elsevier.

Churchill, R. V. (1972). *Operational mathematics*. McGraw-Hill.

Di Benedetto, H., de La Roche, C., Baaj, H., Pronk, A., & Lundström, R. (2004). Fatigue of bituminous mixtures. *Materials and Structures, 37*(3), 202–216. Available from https://doi.org/10.1007/BF02481620.

Fabrizio, M., & Morro, A. (1992). *Mathematical problems in linear viscoelasticity*. SIAM.

Ferry, J. D. (1980). *Viscoelastic properties of polymers*. John Wiley & Sons.

Fini, E. H., Khodaii, A., & Hajikarimi, P. (2016). Fractional viscoelastic study of low-temperature characteristics of biomodified asphalt binders. *Journal of Materials in Civil Engineering, 28*(9), 04016078. Available from https://doi.org/10.1061/(ASCE)MT.1943-5533.0001525.

Hajikarimi, P., Aflaki, S., & Hoseini, A. S. (2013). Implementing fractional viscoelastic model to evaluate low temperature characteristics of crumb rubber and gilsonite modified asphalt binders. *Construction and Building Materials, 49*, 682–687. Available from https://doi.org/10.1016/j.conbuildmat.2013.09.001.

Hajikarimi, P., Nejad, F. M., & Aghdam, M. M. (2018). Implementing general power law to interconvert linear viscoelastic functions of modified asphalt binders. *Journal of Transportation Engineering Part B: Pavements, 144*(2). Available from https://doi.org/10.1061/JPEODX.0000038.

Hopkins, I. L., & Hamming, R. W. (1957). On creep and relaxation. *Journal of Applied Physics, 28*(8), 906–909. Available from https://doi.org/10.1063/1.1722885.

Huang, Y. H. (1993). *Pavement analysis and design*. Englewood Cliffs, NJ: Prentice Hall.

Kassem, E., Zachary, C. G., & Masad, E. (2013). Viscoelastic Poisson's ratio of asphalt mixtures. *International Journal of Geomechanics, 13*(2), 162–169. Available from https://doi.org/10.1061/(ASCE)GM.1943-5622.0000199.

Kim, Y. R., & Lee, Y.-C. (1995). Interrelationships among stiffnesses of asphalt aggregate mixtures. *Asphalt Paving Technology, 64*, 575–610.

Kim, Y. R., Lee, C. Y., & Lee, J. H. (1995). Correspondence principle for characterization of asphalt concrete. *Journal of Materials in Civil Engineering*, 7(1), 59–68. Available from https://doi.org/10.1061/(ASCE)0899-1561(1995)7:1(59).

Kim, Y. R., Lee, H.-J., & Little, D. N. (1997). Fatigue characterization of asphalt concrete using viscoelasticity and continuum damage theory (with discussion). *Journal of the Association of Asphalt Paving Technologists, 66.*

Lee, S. H., & Kim, J. (2009). Determination of viscoelastic Poisson's ratio and creep compliance from the indirect tension test. *Journal of Materials in Civil Engineering, 21*(8), 416–425. Available from https://doi.org/10.1061/(ASCE)0899-1561(2009)21:8(416).

Lindelöf, P., Said, S., & Ahmed, A. (2018). Poisson's ratio of asphalt concrete mixes using indirect tensile test. *Journal of Testing and Evaluation, 47*(1), 498–510. Available from https://doi.org/10.1520/JTE20170516.

Mallela, J., Glover, L. T., Darter, M. I., Von Quintus, H., Gotlif, A., Stanley, M., & Sadasivam, S. (2009). *Guidelines for implementing NCHRP 1-37A ME design procedures in Ohio: Volume 1—Summary of findings, implementation plan, and next steps*. Ohio: Department of Transportation.

Mead, D. W. (1994). Numerical interconversion of linear viscoelastic material functions. *Journal of Rheology, 38*(6), 1769–1795. Available from https://doi.org/10.1122/1.550526.
Ottosen, N. S., & Ristinmaa, M. (2005). *The mechanics of constitutive modeling*. Elsevier.
Park, S. W., & Kim, Y. R. (1999). Interconversion between relaxation modulus and creep compliance for viscoelastic solids. *Journal of Materials in Civil Engineering, 11*(1), 76–82. Available from https://doi.org/10.1061/(ASCE)0899-1561(1999)11:1(76).
Park, S. W., & Schapery, R. A. (1999). Methods of interconversion between linear viscoelastic material functions. Part I—a numerical method based on Prony series. *International Journal of Solids and Structures, 36*(11), 1653–1675. Available from https://doi.org/10.1016/S0020-7683(98)00055-9.
Phan-Thien, N., & Rossikhin, Y. A. (2004). Understanding viscoelasticity: Basics of rheology. *Applied Mechanics Reviews, 57*(1), B4.
Sadd, M. H. (2009). *Elasticity: Theory, applications, and numerics*. Academic Press.
Schapery, R. (1962). Approximate methods of transform inversion for viscoelastic stress analysis. *Proceedings of the Fourth U.S. National Congress of Applied Mechanics, 2*.
Schapery, R. A., & Park, S. W. (1999). Methods of interconversion between linear viscoelastic material functions. Part II—an approximate analytical method. *International Journal of Solids and Structures, 36*(11), 1677–1699. Available from https://doi.org/10.1016/S0020-7683(98)00060-2.
Sorvari, J., & Malinen, M. (2007). Numerical interconversion between linear viscoelastic material functions with regularization. *International Journal of Solids and Structures, 44*(3), 1291–1303. Available from https://doi.org/10.1016/j.ijsolstr.2006.06.029.
Tschoegl, N. W. (2012). *The phenomenological theory of linear viscoelastic behavior: An introduction*. Springer Science & Business Media.
Underwood, B., Far, M., & Kim, Y. (2010). Using limited purchase specification tests to perform full linear viscoelastic characterization of asphalt binder. *Journal of Testing and Evaluation, 38*(5), 558–566. Available from https://doi.org/10.1520/JTE102591.
Witczak, M.W., & Mirza, M.W. (1999). *Development of relationships to predict Poisson's ratio for paving materials* (Interteam Technical Report No. NCHRP 1-37A). College Park, MD: University of Maryland.
Zak, J., Stastna, J., Vavricka, J., Milackova, K., Kasek, L., & Zanzotto, L. (2015). Poisson's ratio of hot asphalt mixtures determined by relaxation and small amplitude oscillation test. *Journal of Testing and Evaluation, 43*(2), 1–7. Available from https://doi.org/10.1520/JTE20130325.
Zhang, W., Drescher, A., & Newcomb, D. E. (1997). Viscoelastic analysis of diametral compression of asphalt concrete. *Journal of Engineering Mechanics, 123*(6), 596–603. Available from https://doi.org/10.1061/(ASCE)0733-9399(1997)123:6(596).

CHAPTER 7

Application of viscoelasticity for experimental tests

7.1 Introduction

Observation and experience are vital in science and are known as the two main tools to explore different aspects of natural phenomena (Shapere, 1982). Scientists use observation to gather and document data, which empowers them to construct theorems and hypotheses and then test them. Many ways can be used to observe by scientists: their own senses or with designed and developed tools such as telescopes, scanners, microscopes, or transmitters to amplify their vision or hearing. Implementing such devices for precise and prearranged observations to find an answer for a specific question is called an "experiment" in science (Gooding et al., 1989). For example, in fluid statics, Abu Rayhan Biruni (973–1048) revealed that the specific gravity of an object has a correlation with the volume of water it replaces. In this manner, he determined the differences in weight between freshwater and salty water, and also between hot and cold water during a scientific experimental procedure by introducing the method of checking tests (Kamiar, 2008). There are several other well-known examples that are not mentioned here for the sake of brevity and falling into the main issue of this chapter.

Performing experiments is one of the main parts of engineering to characterize different types of materials and predict their rheological and mechanical behavior under different loading and boundary conditions. Therefore, lots of experimental methods have been developed to determine constitutive functions (relaxation modulus and creep compliance) of viscoelastic materials as well as other desired properties in both the time and frequency domains like complex shear modulus (G^*), phase angle (δ), storage modulus (G'), loss modulus (G''), dissipated energy, etc. In this chapter, some well-known experimental tests are introduced, focusing on viscoelastic relations required to analyze their results as well as some recently developed methods. These test methods can be categorized based on the test setup and the device used for running the test. For example, it

Applications of Viscoelasticity
DOI: https://doi.org/10.1016/B978-0-12-821210-3.00001-2

is possible to consider the three-point bending beam as a simple test setup which can be performed by using either a bending beam rheometer (BBR) device based on ASTM D6648 for bitumen specimens or a universal testing machine (UTM) for bituminous composites like asphalt mastic or asphalt mixture.

It is evident that there is no chance to present all experimental tests existing for considering the constitutive characteristics of viscoelastic materials. In this chapter, four different simple and widely used test setups are introduced, including three-point bending test, dynamic shear rheometery, rotational viscometer, and UTM. The BBR test and semicircular bending (SCB) test are conducted by using three-point bending test setups. Several well-known modes of loading and boundary conditions can be performed by using a dynamic shear rheometer (DSR) including frequency sweep test, temperature sweep test, time sweep test, creep and recovery, repeated creep and recovery (RCR), multiple stress creep and recovery (MSCR), linear amplitude sweep (LAS) test, and special test setups for capturing healing characteristics of bituminous composites. These different test setups were introduced to assess different aspects of mechanical and rheological behavior of bitumens including rutting susceptibility, resistance to fatigue damages, and elastic recovery properties. The rotational viscometer is utilized to determine the viscosity of original and modified bitumen at different temperatures and the shear rate which is so important for calculating the mixing and compacting temperature of asphalt mixture as well as realizing the behavior of bituminous materials under different loading.

7.2 Three-point bending test

The three-point bending beam is an appropriate setup that can be simply arranged in the laboratory for a wide range of dimensions and loading magnitude under controlled test temperatures. Having the analytical solution in the elastic domain for this setup provides an appropriate basis to find its viscoelastic solution based on the correspondence principle, as it is mentioned in Section 4.3.1. It is also possible to use this simple test setup for primary experiments to evaluate fracture mechanics properties of bituminous composites. In this section, the BBR and SCB tests are presented as two experiments used widely to evaluate the low-temperature characteristics of bitumen and asphalt mixture, respectively.

7.2.1 Bending beam rheometer

The BBR is used to evaluate the low-temperature creep response of the original and modified bitumens. Anderson et al. (1990) at the Pennsylvania State University developed and introduced the bending beam device as part of a federal highway administration (FHWA) project. Then, a new version of this device was developed by Bahia et al. (1992), which is completely computerized and can give an explicit, accurate measurement of the creep properties of bitumens within a low-temperature range. To perform this test, a load of 980 ± 50 mN is applied on a prismatic specimen (127 mm length, 12.7 mm width, and 6.35 mm height), and deflections at the mid-span are measured at time intervals of 0.5 s during 240 s. Fig. 7.1 shows a BBR test specimen and molds used to fabricate it as well as the BBR test setup under loading. Flexural creep stiffness, $S(t)$, and creep rate, m-value, are two primary outputs of the BBR test according to ASTM D6648-08 (2001).

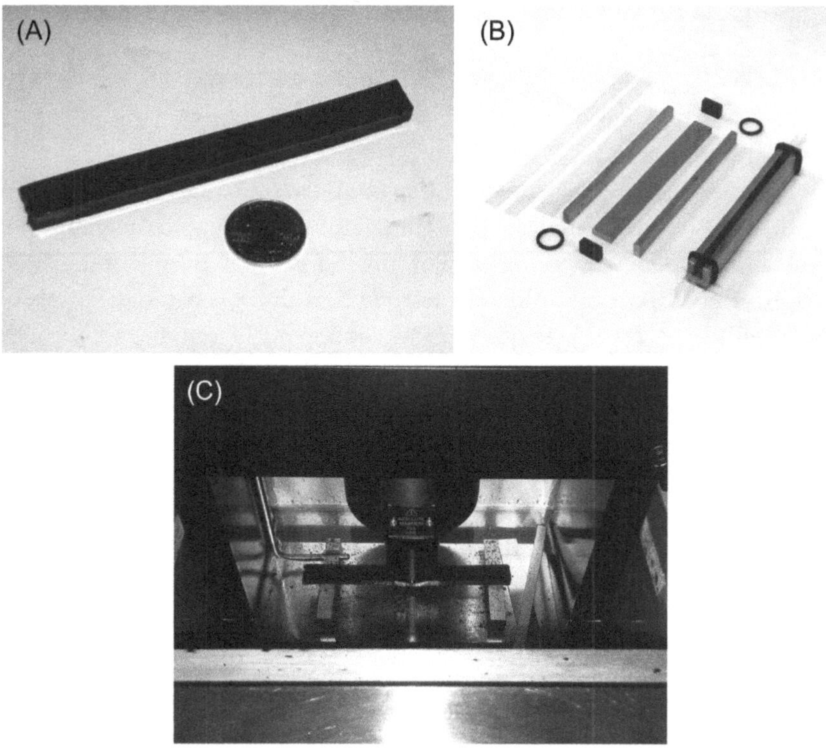

Figure 7.1 The bending beam rheometer test (A) specimen, (B) molds used to fabricate specimens, and (C) test setup.

For an elastic material in a three-point bending setup, the maximum deflection occurred at the mid-span and can be determined by using elementary bending theory (Timoshenko & Young, 1962):

$$\delta = \frac{PL^3}{48EI} \tag{7.1}$$

in which δ is deflection of beam at mid-span (mm), P is applied constant load (N), L is span length (mm), E is modulus of elasticity, I is moment of inertia of the section $= bh^3/12$ (mm^4), b is width of the beam, and h is depth of the beam.

Therefore considering the correspondence principle, the flexural creep stiffness can be defined as follows (see Section 4.3.1):

$$S(t) = \frac{PL^3}{4bh^3\delta(t)} \tag{7.2}$$

The creep rate, m-value, is the absolute value of the slope of the $\log[S(t)] - \log(t)$ curve determined by using the following equation:

$$m = \frac{d\log[S(t)]}{d\log(t)} \tag{7.3}$$

in which t is time, s.

The flexural creep stiffness, $S(t = 60\text{ s})$, and m-value ($t = 60$ s) are simultaneously used to determine the critical cracking temperature of original and modified bitumens. For this purpose, it is conventional to perform the BBR test at different test temperatures. Test temperatures of 0°C, −6°C, −12°C, −18°C, −24°C, and −30°C can be appropriate alternatives. However, there is no conceptual limitation for selecting other test temperatures. The critical cracking temperature was defined as the higher of the two temperatures corresponding to a $S(t = 60\text{ s})$ of 300 MPa and an m-value ($t = 60$ s) of 0.3 based on ASTM D6648-08 (2001). Fig. 7.2 depicts the results of a BBR test performed on bitumen modified with 1.3% of polyphosphoric acid (PPA) at different test temperatures (Hajikarimi et al., 2013). Based on the mid-span deflection captured during the test, the flexural creep stiffness and m-value can be simply calculated at $t = 60$ s to check the standard criteria to determine critical cracking temperature. Fig. 7.3 shows both $S(t)$ and m-value considering the maximum criteria [$S(t = 60\text{ s}) < 300$ MPa] for flexural creep stiffness and the minimum criteria for m-value [$m(t = 60\text{ s}) > 0.3$]. As can be seen in this figure, the criterion for the maximum amount of $S(t)$ did not pass

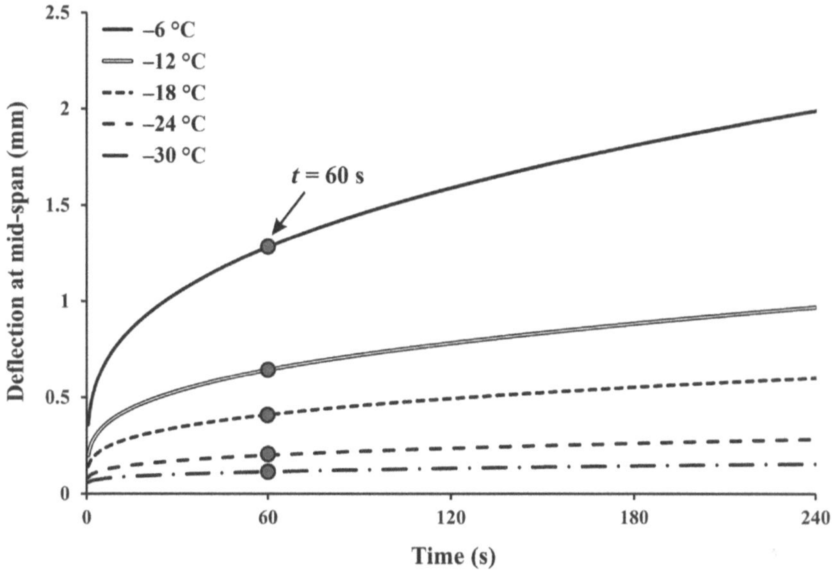

Figure 7.2 Deflection at mid-span for a bending beam rheometer test on bitumen modified with 1.3% of polyphosphoric acid at five different test temperatures.

at test temperatures of −30°C and −24°C. On the other hand, the criterion for the minimum amount of *m*-value did not pass at test temperatures of −30°C, −24°C, −18°C, and −12°C. Thus, the critical cracking temperature is determined as −17.7°C by adding −10°C to the highest real temperature interpolated to pass both criteria.

In addition to conventional analysis on the BBR test data, it is possible to use viscoelastic models to further analyze and extract more useful information. Hajikarimi et al., (2018b) determined the relaxation modulus of modified bitumen at low temperatures based on creep data acquired from the BBR test by using interconversion methods including the Hopkins and Hamming method and fractional viscoelastic modeling (see Chapter 4: Correspondence Principle of Viscoelasticity). Here, another use of creep data is presented by introducing a new index of "derivation of creep compliance" derived by Liu et al. (2010) to evaluate low-temperature performance of original and modified bitumens.

In log–log scale, the changing trend of $\log[S(t)]$ over $\log(t)$ can be approximated by using a second-order parabola as:

$$\log[S(t)] = A + B[\log(t)] + C[\log(t)]^2 \qquad (7.4)$$

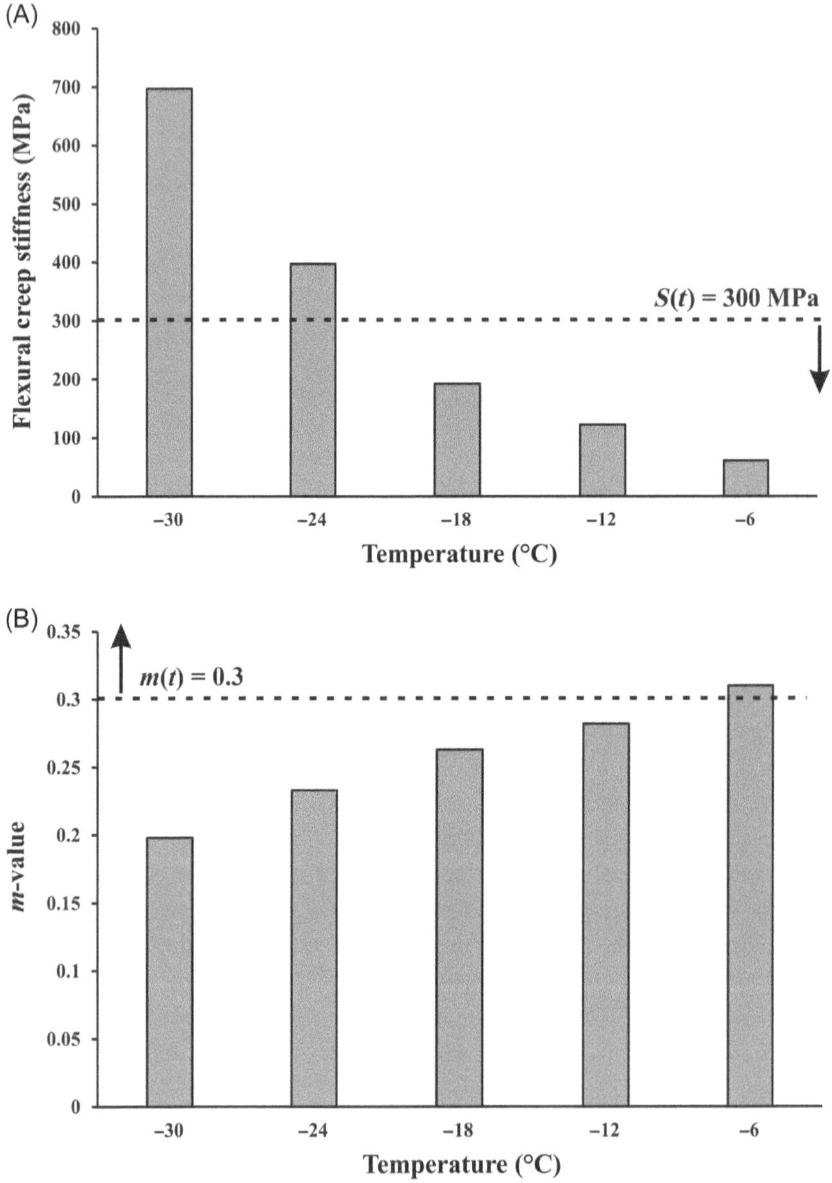

Figure 7.3 Parameters determined based on results of the bending beam rheometer test on bitumen modified with 1.3% of polyphosphoric acid: (A) flexural creep stiffness and (B) *m*-value at $t = 60$ s.

in which A, B, and C are regression coefficients. Fig. 7.4 shows a schematic curve of $\log[S(t)]$ versus $\log(t)$. Regarding Eq. (7.3), the creep rate, $m(t)$, can be calculated as:

$$m(t) = \left|\frac{d\{\log[S(t)]\}}{d[\log(t)]}\right| = |B + 2C[\log(t)]| \tag{7.5}$$

Considering Eq. (7.4), the following equation can be simply derived:

$$S(t) = 10^{\{A+B[\log(t)]+C[\log(t)]^2\}} \tag{7.6}$$

Let

$$M_1(t) = \frac{d[S(t)]}{dt} \tag{7.7}$$

Eq. (7.7) can be expanded by substituting Eq. (7.6) into this equation as follows:

$$M_1(t) = 10^{\{A+B[\log(t)]+C[\log(t)]^2\}} \times [B + 2C\log(t)]/t \tag{7.8}$$

Regarding the properties of $\log[S(t)]$ in the log–log coordinate system as shown in Fig. 7.4, it can be found that:

$$B + 2C\log(t) < 0 \tag{7.9}$$

Therefore it can be concluded that:

$$B + 2C\log(t) = -m(t) \tag{7.10}$$

Using Eqs. (7.6) and (7.10) for terms mentioned in Eq. (7.8), the $M_1(t)$ can be written as:

$$M_1(t) = \frac{-S(t) \times m(t)}{t} \tag{7.11}$$

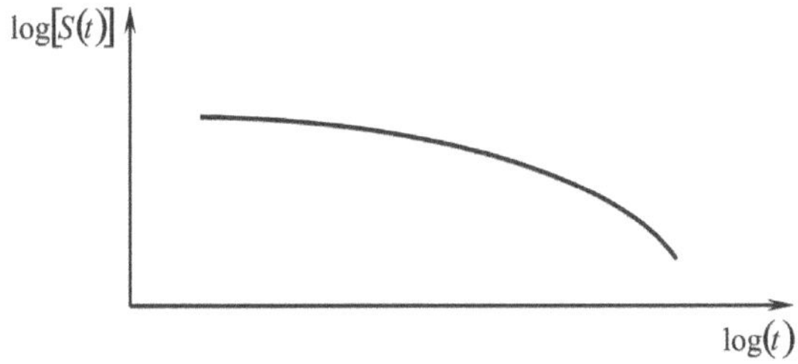

Figure 7.4 Schematic curve of log[$S(t)$] versus log(t) for the bending beam rheometer test.

On the other hand, flexural creep stiffness, $S(t)$, is intrinsically defined as the inverse of the creep compliance, $D(t)$, and can be obtained by the following equation:

$$S(t) = \frac{1}{D(t)} \tag{7.12}$$

Again, let

$$M_2(t) = \frac{d[S(t)]}{dt} \tag{7.13}$$

The Burgers model which is introduced in Chapter 3, Mechanical Models of Viscoelasticity, can be considered to reflect the mechanical behavior of original and modified bitumens at low temperatures. Referring to the parameters illustrated in Fig. 3.7, the creep compliance is explained as:

$$D(t) = \frac{1}{E_M} + \frac{t}{\eta_M} + \frac{1}{E_K}\left(1 - e^{-E_K t/\eta_K}\right) \tag{7.14}$$

By substituting Eqs. (7.14) and (7.12) into Eq. (7.13), the following will be determined:

$$M_2(t) = -\frac{\frac{1}{\eta_M} + \frac{1}{\eta_K} e^{-\frac{E_K}{\eta_K}t}}{\left(\frac{1}{E_M} + \frac{t}{\eta_M} + \frac{1}{E_K}\left(1 - e^{-\frac{E_K}{\eta_K}t}\right)\right)^2} \tag{7.15}$$

It is evident that $M_1(t)$ and $M_2(t)$ have the equal physical meaning and the following equality can be written:

$$\frac{-S(t) \times m(t)}{t} \approx -[S(t)]^2 \times \left(\frac{1}{\eta_M} + \frac{1}{\eta_K} e^{-\frac{E_K}{\eta_K}t}\right) \tag{7.16}$$

which is concluded as follows:

$$\frac{m(t)}{S(t)} \times \frac{1}{t} \approx D'(t) \tag{7.17}$$

where $D'(t)$ is derivation of creep compliance as:

$$D'(t) = \frac{1}{\eta_M} + \frac{1}{\eta_K} e^{-\frac{E_K}{\eta_K}t} \tag{7.18}$$

Consequently, at a specific time, Eq. (7.17) demonstrates that increasing the amount of $m(t)$ and decreasing the $S(t)$ simultaneously leads to a higher derivation of creep compliance, $D'(t)$. According to ASTM D6648 and the aforementioned criteria for both $S(t)$ and $m(t)$, these conditions cause a better performance for bituminous materials. Therefore, $D'(t)$ can be introduced as a performance index for original and modified bitumens at low temperatures. The higher the amount of $D'(t)$, the lower low-temperature cracking.

Aflaki and Hajikarimi (2012) used this index to rank-modified bitumen with different types of additives including crumb rubber, styrene–butadiene–styrene (SBS), and PPA in terms of their anticracking properties at low temperatures based on the BBR test results. They showed that the relative ranking of modified bitumen does not change by substituting various amounts of t into the derivation of the creep compliance equation. For example, Fig. 7.5 shows $D'(t)$ versus test temperature for crumb rubber–modified bitumen at 60 s, 240 s, and $t \to \infty$. It can be seen in this figure that greater time causes the curves to shift up but the order of curves does not change.

7.2.2 Semicircular bending test

There are several fracture tests which can be used to investigate fracture behavior of viscoelastic substances like bituminous materials including the single-edge notched beam test (Hoare & Hesp, 2000; Mobasher et al., 1997; Wagoner et al., 2005), the disk-shaped compact tension test (Wagnoner et al., 2005, 2006), the double-edged notched tension test (Youngguk, 2003), and the SCB test (Marasteanu et al., 2007; Molenaar et al., 2002). The SCB has attracted more attention during the last years, as having a low amount of asphalt mixture it will be possible to conduct several tests. Fig. 7.6 depicts two well-known setups of the SCB test to investigate the fracture behavior of asphalt mixture in mode I and mixed-mode of fracture by using a vertical and inclined notch. This figure clearly shows that the SCB test has a three-point bending setup with a couple of simple supports and a concentrated load at its center. The test can be performed by applying a specified constant or variable force/displacement.

The main output of the SCB test is the force–displacement curve, depicted in Fig. 7.7 for an asphalt mixture at different test temperatures including −10°C, 0°C, and 10°C. As can be seen in this figure, there are two major points on this curve: the point of maximum force (peak load) shown with label "P" in Fig. 7.7 and the point at the end of the curve shown with label "D" where the measured force reaches to 5% of the

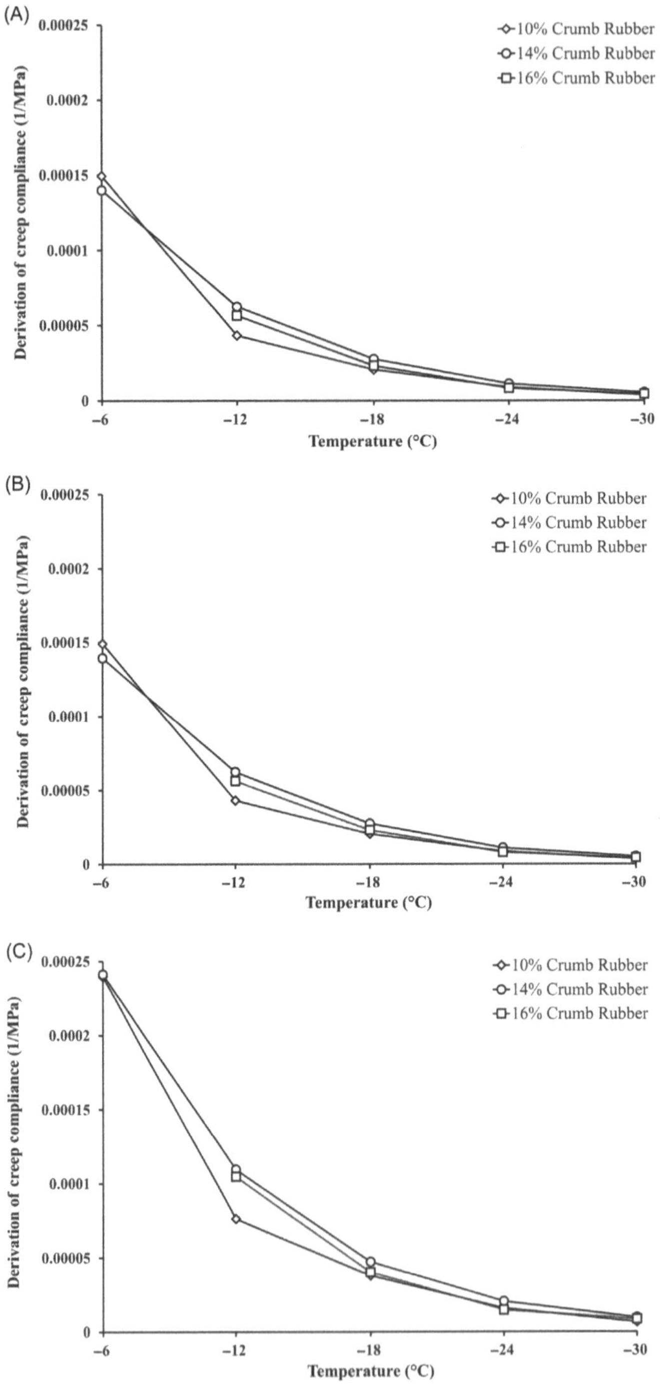

Figure 7.5 Derivation of creep compliance for crumb rubber–modified bitumens versus test temperature for (A) 60 s, (B) 240 s, and (C) infinite time. *From Aflaki, S., & Hajikarimi, P. (2012). Implementing viscoelastic rheological methods to evaluate low temperature performance of modified asphalt binders. Construction and Building Materials, 36, 110–118.*

(A)
(B)
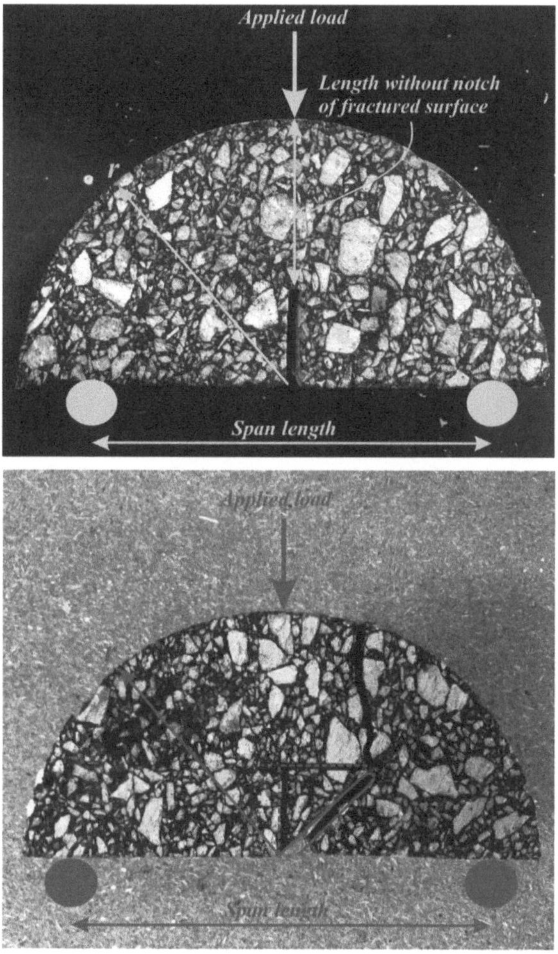

Figure 7.6 Two well-known setups of the SCB test for (A) mode I and (B) mixed mode of fracture.

peak load. The higher amount of peak load shows that the tested specimen has a great resistance against fracture. After reaching the peak load, the prescribed crack designated within the specimen initiates to propagate and if the second point associates with greater displacement, it can be concluded that the specimen has higher ductility and is able to deform more after crack propagation.

According to a force–displacement curve captured during a SCB test, it is possible to determine fracture energy and fracture toughness. In the following sections, both properties are briefly introduced.

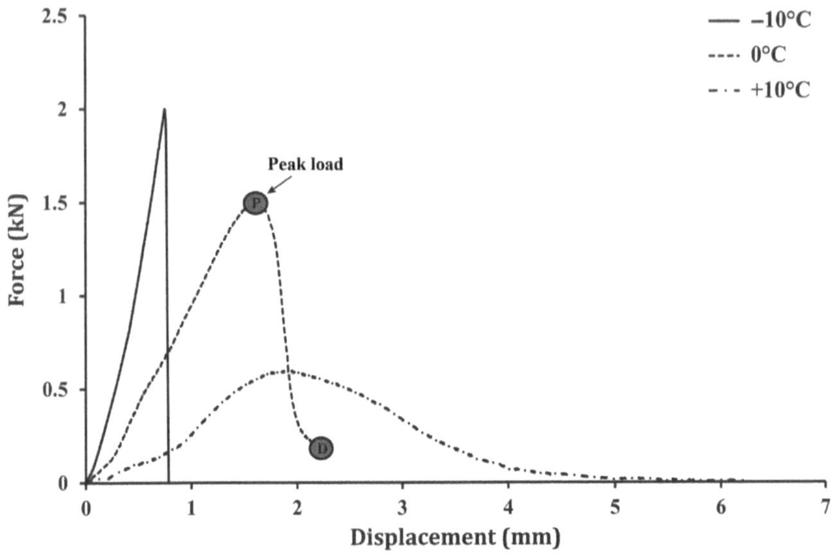

Figure 7.7 Force–displacement curve as the main output of the semicircular bending test for an asphalt mixture at three different test temperature in mode I of fracture.

7.2.2.1 Fracture energy

The fracture energy can be calculated by dividing the work done for performing the fracture process (the area under the force–displacement curve) by the fractured zone area. The area of fractured zone is determined by multiplying the thickness of the specimen by the length of the portion without having a notch, as shown in Fig. 7.8. To formulate the abovementioned descriptions, the following equation can be used:

$$G_f = \frac{W_f}{A_{lig}} \tag{7.19}$$

where G_f is fracture energy (J/m^2); W_f is the work done to perform fracture process which is $W_f = \int P.du$; P is applied load (N); u is displacement in line with force (m); r is specimen's radius (m); a is notch length (m); and t is specimen's thickness (m).

The area under the force–displacement curve can be calculated by using the quadrangle rule as follows:

$$W = \sum_{i=1}^{n} (u_{i+1} - u_i) \times P_i + \frac{1}{2}(u_{i+1} - u_i)(P_{i+1} - P_i) \tag{7.20}$$

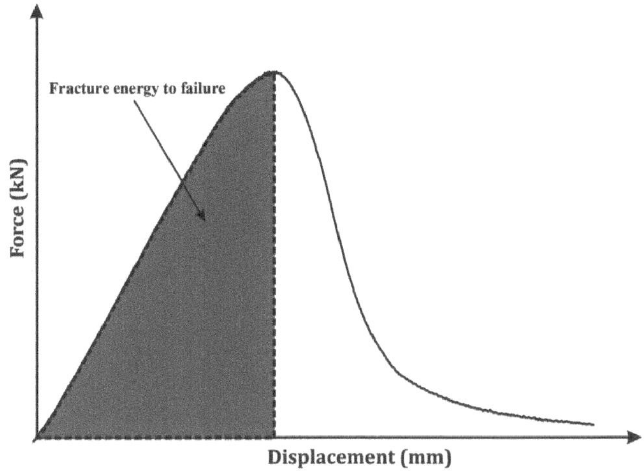

Figure 7.8 Fracture energy to failure and fracture energy in force–displacement curve.

in which P_i is applied force (N) at ith stage of loading; P_{i+1} is applied force (N) at $i+1$th stage of loading; u_i is displacement in line with applied force (m) at ith stage of loading; and u_{i+1} is displacement in line with applied force (m) at $i+1$th stage of loading.

As can be seen in Fig. 7.7, increasing the test temperature reduces peak load and extends terminal displacement, known as ductility, during the fracture process. Therefore it is possible to achieve the same fracture energy for two specimens which have different peak loads and different terminal displacements.

7.2.2.2 Fracture toughness

The fracture toughness is the product of stress intensity factor (SIF), K_I, at critical load P_c. The following equation can be used for calculating the SIF of mode I of fracture (Li & Marasteanu, 2004):

$$\frac{K_I}{\sigma_0\sqrt{\pi a}} = Y_{I(0.8)} \tag{7.21}$$

in which $\sigma_0 = P/2rt$; P is applied load (MN); r is specimen's radius (m); t is specimen's thickness; a is notch length (m); and Y_I is normalized SIF (dimensionless) which can be determined as follows:

$$Y_{I(0.8)} = 4.782 + 1.219\left(\frac{a}{r}\right) + 0.063\exp\left(7.045\left(\frac{a}{r}\right)\right) \tag{7.22}$$

Using the peak load of the force–displacement curve as P_c, it will be possible to determine fracture toughness.

7.3 Dynamic shear rheometer

One of the most important experimental devices developed and used to characterize rheological and mechanical properties of viscoelastic materials at different test temperatures and loading rates is DSR (Eschbach, 1993). The DSR is utilized in different fields of research and practical studies including polymer engineering, chemical engineering and pavement engineering. An applicable manifestation of using DSR to characterize viscoelastic materials is performance grading (PG) of asphalt binders according to the superpave protocol for both original and modified asphalt binders (Bahia et al., 2001). The actual pavement high/low temperatures expected based on climatic conditions are determined in the area where the pavement will be placed. Then, an asphalt binder will be selected for such a climate condition having an appropriate PG (higher and lower supportable temperatures). For example, for an area with probable higher and lower pavement temperature of 57°C and −20°C, a PG58-22 asphalt binder can be used to meet superpave criteria.

The simple DSR test uses a thin bitumen specimen (see Fig. 1.7) placed between two circular plates. The lower plate is fixed while the upper plate oscillates back and forth across the sample at a frequency of 10 rad/s (1.59 Hz) to apply a torsional force which appears as shear stress on the specimen's perimeter (see Fig. 7.9). To consider all aging

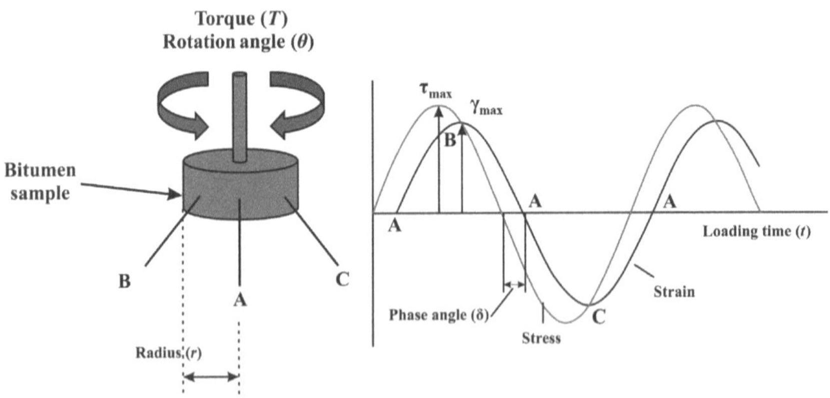

Figure 7.9 Dynamic shear rheometer test setup for applying a harmonic exciting shear stress and capturing response strain.

conditions of bitumens after production, mixing, and placing, the DSR tests are conducted on unaged, short-term aged by using rolling thin film oven (RTFO) test (ASTM D175M-09, 2014) and long-term aged samples by using pressure aged vessel (PAV) (ASTM D6521-19a, 2019) test. It is necessary to mention that, as depicted in Fig. 1.8, there are two different DSR test setups: parallel plate and cone plate.

In the superpave protocol, the DSR test is used to investigate viscoelastic properties of bitumens at high and moderate test temperatures to evaluate rutting and fatigue susceptibility of a bitumen, respectively. Since a bitumen sample has the most critical condition to resist against permanent deformation at early ages, the short-term aged samples are used to investigate high-temperature characteristics of bitumens. On the other hand, fatigue performance of the pavement deteriorates with the passing of time and long-term aging of bituminous materials. Therefore the long-term aged samples should be used to evaluate the moderate temperature behavior of bitumens.

The work performed is partially recovered by the elastic component of the strain and the remaining is dissipated by its viscous flow component and any related generation of heat throughout the deformation of bitumen at high temperatures. Anderson et al. (1994) assumed that permanent deformation is produced by the total energy dissipated per cycle of loading and can be calculated as:

$$\Delta U = \int \tau \, d\gamma \tag{7.23}$$

For a sine wave loading from 0 to 2π, Eq. (7.23) can be calculated as:

$$\Delta U = \pi \, \tau_{max} \, \gamma_{max} \sin\delta \tag{7.24}$$

in which ΔU is energy loss per cycle or dissipated energy; τ_{max} is maximum shear stress; γ_{max} is maximum shear strain; and δ is phase angle. Since $G^* = \tau_{max}/\gamma_{max}$, Eq. (7.24) can be rewritten as:

$$\Delta U = \pi \, {\tau_{max}}^2 \frac{1}{\frac{G*}{\sin\delta}} \tag{7.25}$$

As can be seen in this equation, $G^*/\sin\delta$ is inversely relative to the dissipated energy by the viscous flow component of the strain. Increasing $G^*/\sin\delta$ decreases the total dissipated energy and rutting vulnerability will be reduced. Therefore the $G^*/\sin\delta$ was identified as a superpave specification parameter for evaluating high-temperature PG of asphalt binders.

According to superpave protocol, the high service temperature of a bitumen is determined as the temperature at which $G^* / \sin\delta$ is greater than 1 kPa for an unaged sample and greater than 2.2 kPa for the short-term aged condition by using the RTFO test.

To resist against fatigue cracking, a bitumen sample should be elastic (capable to dissipate energy without cracking) but not too stiff. Thus a minimum complex shear modulus $G^* \sin\delta$ should be met. The fatigue cracking can be considered a stress-controlled phenomenon in thick hot mix asphalt (HMA) pavements and a strain-controlled phenomenon in thin HMA pavements. Since fatigue cracking is more predominant in thin pavements, the strain-controlled test is of most concern to investigate fatigue resistance of bitumens. During each cycle of traffic loading as the main source of fatigue cracking, work is being done to deform the pavement surface. As mentioned before, part of this work is recovered by the elastic rebound of the pavement surface, while part is dissipated in the form of rutting, heat, cracking, and crack propagation. The lower the amount of energy dissipated per loading cycle the less likely fatigue cracking is. Therefore to minimize fatigue cracking, it is required to minimize the amount of work dissipated per loading cycle which is as follows at a constant strain:

$$W = \pi \varepsilon_0^2 [G^* \sin\delta] \tag{7.26}$$

in which W is work dissipated per load cycle; ε_0 is strain during load cycle; G^* is complex modulus; and δ is phase angle

To mitigate fatigue cracking, there is a limiting maximum stiffness at intermediate temperatures based on superpave protocol. The intermediate service temperature is equal to [High service temperature + Low service temperature]/2 at which $G^* \sin\delta$ in the PAV aged condition does not exceed 5000 kPa.

In addition to the regular DSR test to evaluate PG grading of bitumens at frequency of 1.59 Hz, there are different modes to capture the rheological and mechanical behavior of viscoelastic material for a specific purpose. In the following, these modes are introduced, including frequency sweep test, temperature sweep test, time sweep test, creep and recovery, RCR, MSCR, LAS test, and healing characterization test.

7.3.1 Frequency sweep test

The dynamic oscillatory shear test is generally conducted within the region of linear viscoelastic response (LVE). This test is performed to measure the complex shear modulus (G^*) and phase angle (δ) which mutually describe the viscoelastic response of bituminous materials. A rheometer under the controlled strain conditions at desired test temperatures can be used to sweep frequencies from the greatest frequency, which is usually 100 Hz, to the lowest, which is usually 0.1 Hz. With regard to the test temperature used to run the frequency sweep test, the specimen dimensions must be considered, as for temperatures higher than 40°C, a 25 mm diameter plate and 1 mm gap opening was used, and for temperatures lower than 40°C, a 8 mm diameter plate and 2 mm gap opening should be used (see Fig. 7.10).

The shear strain for performing the frequency sweep test should guarantee remaining within the linear viscoelastic region during the testing. A shear strain between 0.01% (0.0001 mm/mm) and 0.1% (0.001 mm/mm) can be an appropriate alternative to conduct the frequency sweep test on original and modified bitumens.

In the frequency sweep test, a sinusoidal shear strain is applied in each frequency and the associated shear stress is measured. The sinusoidal shear strain and shear stress can be represented as:

$$\gamma(t) = \gamma_0 e^{i(\omega t - \delta)} \tag{7.27}$$

$$\tau(t) = \tau_0 e^{i\omega t} \tag{7.28}$$

where τ_0 and γ_0 are stress and strain amplitude; ω is angular frequency; δ is phase angle; and t is time. The real and imaginary parts of the complex shear modulus (G^*) can be determined as follows:

$$G^*(\omega) = \frac{\tau_0 e^{i\omega t}}{\gamma_0 e^{i(\omega t - \delta)}} = |G^*| e^{i\delta} = |G^*| \cos\delta + i|G^*| \sin\delta = G' + iG'' \tag{7.29}$$

in which $|G^*|$ is the norm of complex shear modulus, G' is storage modulus (real component of complex shear modulus), and G'' is loss modulus (imaginary component of complex shear modulus) (Kim, 2009).

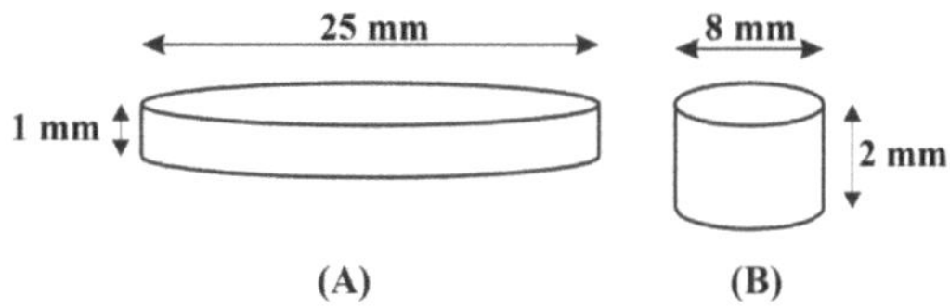

Figure 7.10 Two dynamic shear rheometer specimens used in superpave protocol: (A) 25 mm diameter specimen for performing experiments at test temperatures higher than 40°C, and (B) 8 mm diameter specimen for performing experiment at test temperatures lower than 40°C.

The complex viscosity (η^*) is another important output of the frequency sweep test. It is equal to the difference between the dynamic viscosity and the out-of-phase viscosity, or the imaginary part of the complex viscosity (Shaw & MacKnight, 2018). Fig. 7.11 shows the complex viscosity of crumb rubber–modified bitumens as an example captured during a frequency sweep test (Hajikarimi et al., 2015). Shear-thinning or pseudoplasticity is a phenomenon in which viscosity decreases by increasing the shear rate. This phenomenon was frequently reported for the rheological behavior of modified asphalt binders. At low shear rates, pseudoplastic liquids behave like Newtonian liquids which mean that their viscosity is independent of frequency or shear rate. This viscosity is known as Newtonian viscosity or zero-shear viscosity (ZSV) (Yildirim et al., 2000).

Sybilski (1996) showed that rutting resistance of bituminous mixtures is related to zero-shear viscosity of bitumens. To determine the ZSV by

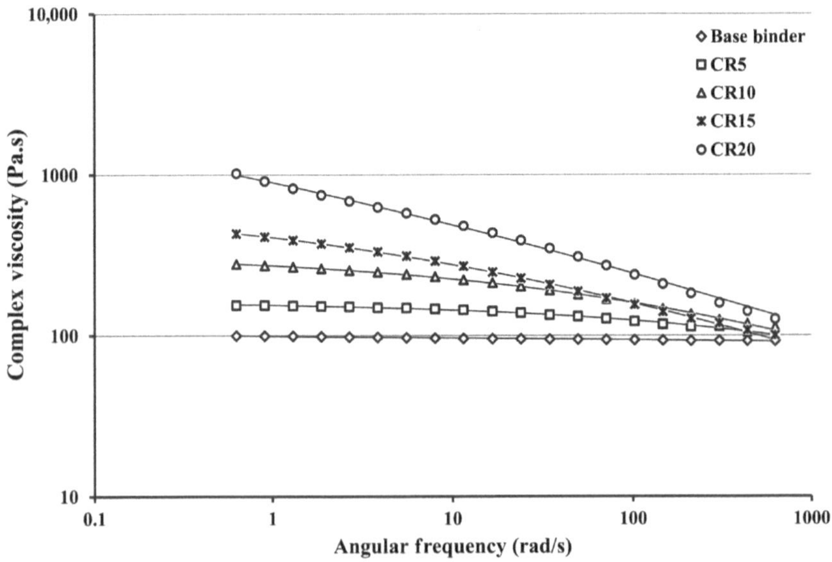

Figure 7.11 Complex viscosity of the base and crumb rubber (CR)–modified bitumens at test temperature of 60°C. *From Hajikarimi, P., Rahi, M., & Moghadas Nejad, F. (2015). Comparing different rutting specification parameters using high temperature characteristics of rubber-modified asphalt binders. Road Materials and Pavement Design, 16(4), 751–766.*

implementing the frequency sweep test outputs, the cross model can be used, which is a four-parameter model as follows (Sybilski, 1994, 1996):

$$\frac{\eta^* - \eta^*_{\infty}}{\eta^*_o - \eta^*_{\infty}} = \frac{1}{1 + (K\omega)^m} \tag{7.30}$$

where η^* is complex viscosity; η^*_o is ZSV; η^*_{∞} is limiting viscosity in the second Newtonian region; ω is angular frequency (rad/s); and K and m are constant parameters.

The frequency sweep test is frequently performed by considering frequencies between 0.1 and 100 Hz. In this frequency range, it is possible to assume that $\eta^* >> \eta^*_{\infty}$. Considering such an assumption, the cross model can be rewritten as (Anderson et al., 2002):

$$\eta^* = \frac{\eta^*_o}{1 + (K\omega)^m} \tag{7.31}$$

Therefore it is possible to fit Eq. (7.31) on complex viscosity versus angular frequency curve by using curve-fitting tools at each test temperature to determine the ZSV. Guericke and Hoeppel (2000) reported that increasing ZSV leads to improving antirutting performance of bitumen.

It is necessary to mention that the frequency sweep test can be used to specify PG grading of a bitumen if the frequency of 1.59 Hz is set as one of the frequency measurement points to capture G^* and δ.

7.3.2 Temperature sweep test

As described in Section 7.3.1, during a frequency sweep test, the test temperature is kept constant and frequency sweeps a prescribed range. It is possible to perform the DSR test at a constant frequency by sweeping the test temperature. For example, Fig. 7.12 depicts complex modulus and phase angle of the nanosilica-modified samples as a function of the nanosilica dosage and temperature at a frequency of 1.59 Hz. As can be seen in this figure, increasing test temperature decreases complex shear modulus for all samples. On the other hand, it is expected that increasing test temperature increases the phase angles, which shows the more viscous behavior of the original and modified bitumens (Fini, Hajikarimi, Rahi, & Nejad, 2016).

7.3.3 Time sweep test

The time sweep test was designed to evaluate fatigue damage implementing the conventional experimental concept as the material integrity deteriorates

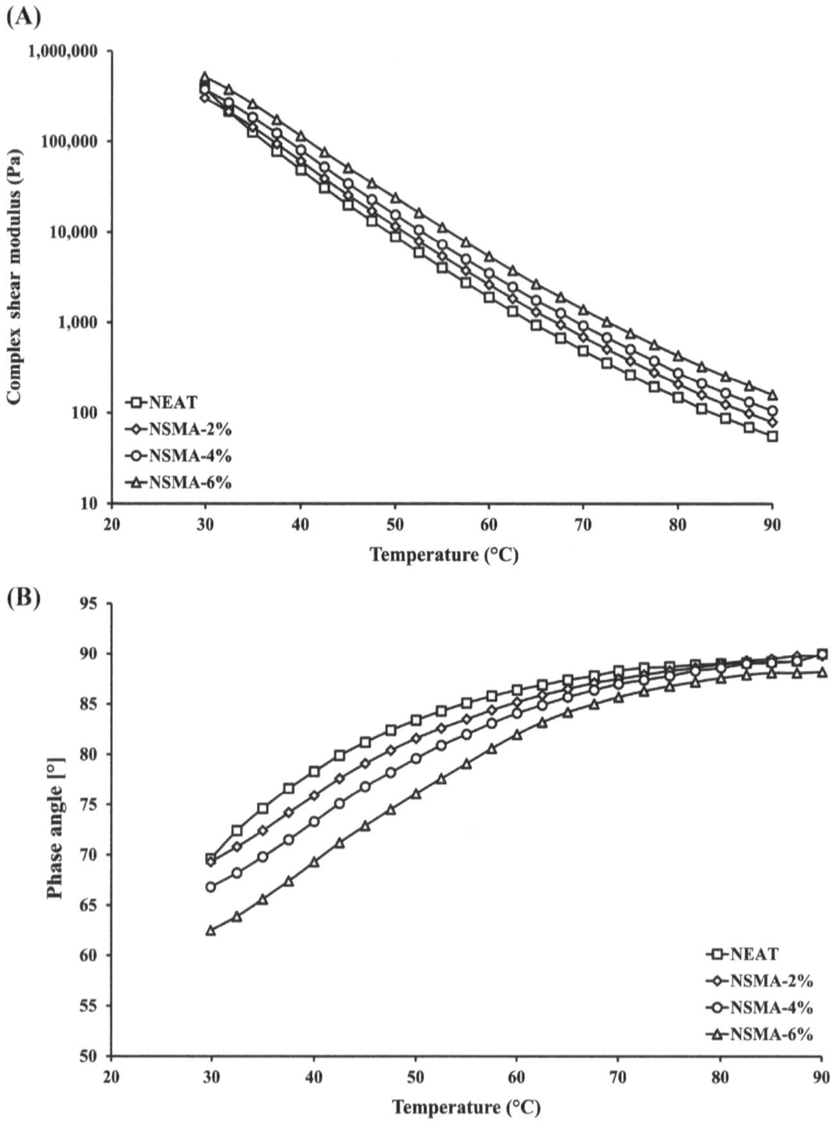

Figure 7.12 Results of temperature sweep test for (A) complex modulus and (B) phase angle of nanosilica-modified samples as a function of the nanosilica dosage and temperature at frequency of 1.59 Hz. NSMA, nano-silica modified asphalt; NEAT is the original asphalt. *From Fini, E. H., Hajikarimi, P., Rahi, M., & Moghadas Nejad, F. (2016). Physiochemical, rheological, and oxidative aging characteristics of asphalt binder in the presence of mesoporous silica nanoparticles. Journal of Materials in Civil Engineering, 28(2), 04015133.*

under repeated loading (Wang et al., 2016). The procedure of the time sweep test consists of applying repeated sinusoidal loading at a specific frequency and amplitude in either controlled displacement or controlled stress mode. Fig. 7.13 shows an example of time sweep test results for the original and styrene–ethylene/propylene–styrene–modified bitumens. Several research works have been devoted to defining an appropriate fatigue failure criteria in bituminous material tests. The simplest fatigue failure definition used in bituminous material tests is 50% loss in stiffness or pseudostiffness (Kim et al., 1997; Tayebali et al., 1992), which can be estimated by using the time sweep test. However, the stiffness reduction approach for defining fatigue failure has been challenged because it is an arbitrary criterion without any phenomenological or theoretical justification.

A DSR with 8-mm parallel plate geometry can be employed to conduct the time sweep fatigue tests on original and modified bitumens. Sketching complex shear modulus versus time resulted in making a sense about occurring fatigue damage in a bitumen and comparing its fatigue behavior with other bitumens. Furthermore, it is possible to use this curve of degrading complex shear modulus within numerical framework in which it is required to involve degradation trend in the numerical analysis.

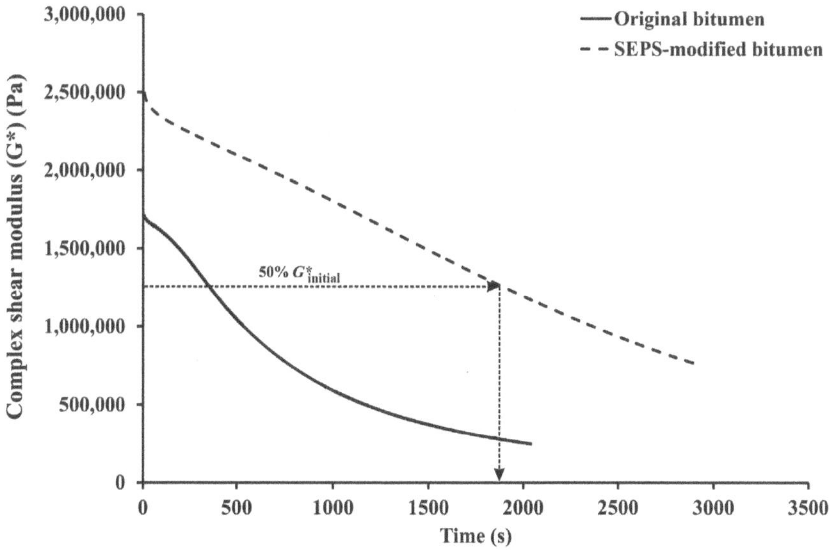

Figure 7.13 Time sweep test results for the original and styrene–ethylene/propylene–styrene (SEPS)–modified bitumens at test temperature of 25°C.

7.3.4 Creep and recovery

The creep and recovery test can be performed by using the DSR to simulate traffic loads on pavement structure. The loading phase can model the tire(s) load followed by the recovery phase which can model the rest period between two passing vehicles. Fig. 7.14 schematically illustrates a creep and recovery test which is conducted by Hajikarimi et al. (2018a,b) to evaluate the nonlinear viscoelastic behavior of the original and modified asphalt binders. They changed loading time (1 and 100 s) and applied stress level (0.1, 1, 10, 20, and 30 kPa) to obtain nonlinear viscoelastic behavior at a constant test temperature of 40°C. The recovery phase for all tests was 999 s, an extremely long time, to ensure full recovery. Fig. 7.15 shows results of creep and recovery test for the original bitumen at different stress levels for creep loading time of 1 and 100 s. As can be seen in this figure, it is not possible to represent experimental data at the stress level of 30kPa as the sample lost deformation resistance during creep loading.

By altering the stress level, the proportion of applied stress to measured strain must remain constant if the viscoelastic material behaves linearly. Hajikarimi et al. (2018a,b) defined the peak strain ratio for each stress level as follows:

$$Peak\ strain\ ratio = \frac{\gamma(\sigma)}{\gamma(\sigma = 0.1\ kPa)} \tag{7.32}$$

in which $\gamma(\sigma)$ is strain at desired stress level and stress ratio is defined as proportion of applied stress to minimum stress level.

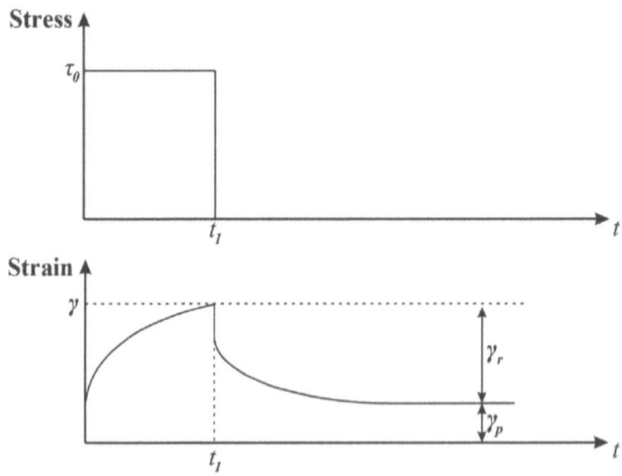

Figure 7.14 Schematic illustration of a creep and recovery test.

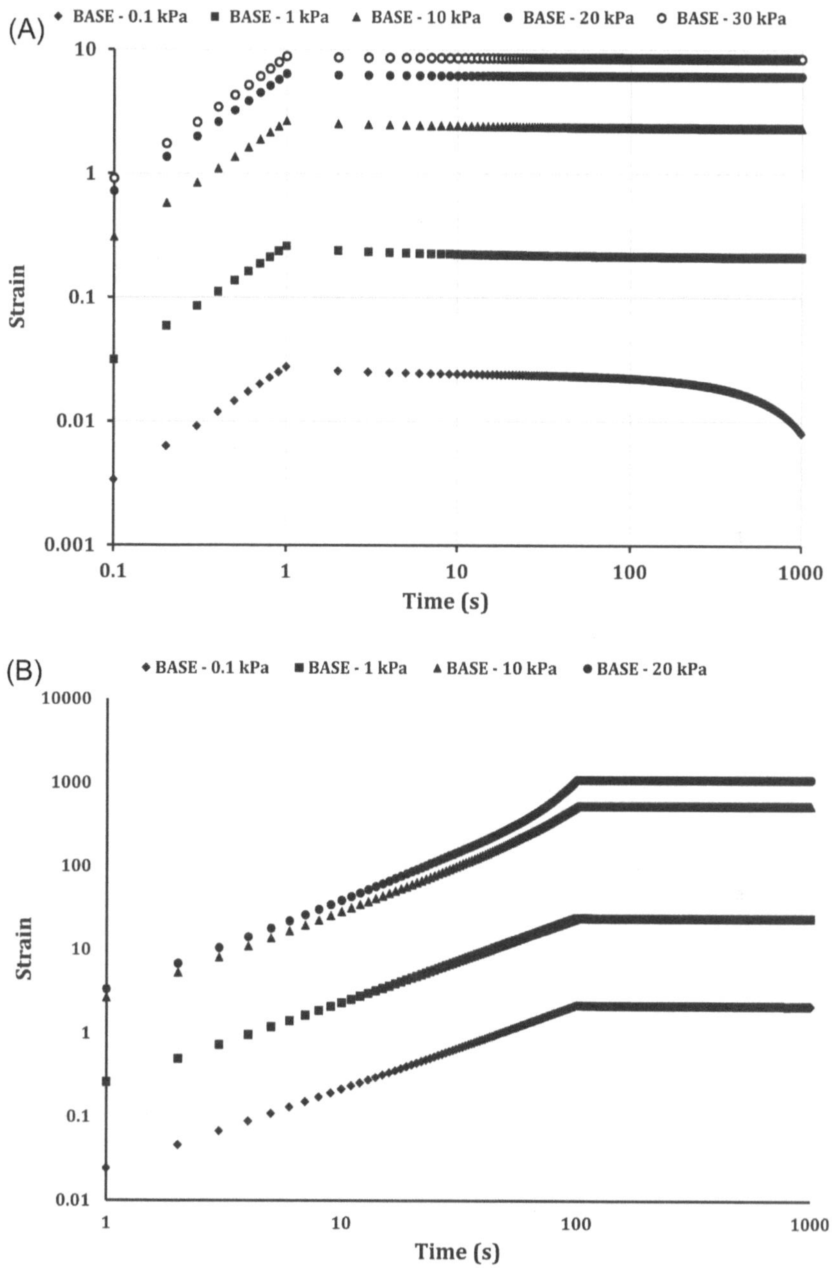

Figure 7.15 Results of creep and recovery test for the original bitumen at different stress levels of 0.1, 1, 10, 20, and 30 kPa for creep loading time of (A) 1 s and (B) 100 s. BASE is the original bitumen without any modifier. *From Hajikarimi, P., Nejad, F. M., Khodaii, A., & Fini, E. H. (2018). Introducing a stress-dependent fractional nonlinear viscoelastic model for modified asphalt binders. Construction and Building Materials, 183, 102–113.*

Then they used this ratio to investigate the effect of bitumen modification by different types of modifiers including crumb rubber, SBS, and PPA on viscoelastic behavior of the original bitumen. For this purpose, a simple figure was used in which peak strain ratio was sketched versus stress ratio. It is evident that for an ideal linear viscoelastic behavior, an identity line ($y = x$) should be captured. Fig. 7.16 depicts peak strain ratio versus stress ratio for both creep loading time of 1 and 100 s and different dosages of crumb rubber–modified bitumen. For example, as can be seen in this figure, adding crumb rubber to the original bitumen causes nonlinearity in viscoelastic behavior since curves related to crumb rubber–modified bitumens are diverged from the line of ideal linear viscoelastic behavior ($y = x$).

Fig. 7.16 obviously depicts the effect of adding bitumen modifiers on nonlinear characteristics of viscoelastic behavior in comparison with the original bitumen. The main shortcoming about this figure is that it is qualitative and just represents the trend of nonlinear behavior by increasing the modifiers' dosages. To quantify the effect of modification on nonlinear viscoelastic properties of bitumen and comparing different types and dosage of modifiers together, a generalized fractional viscoelastic model (GVFM) introduced in Section 3.3.3 can be used.

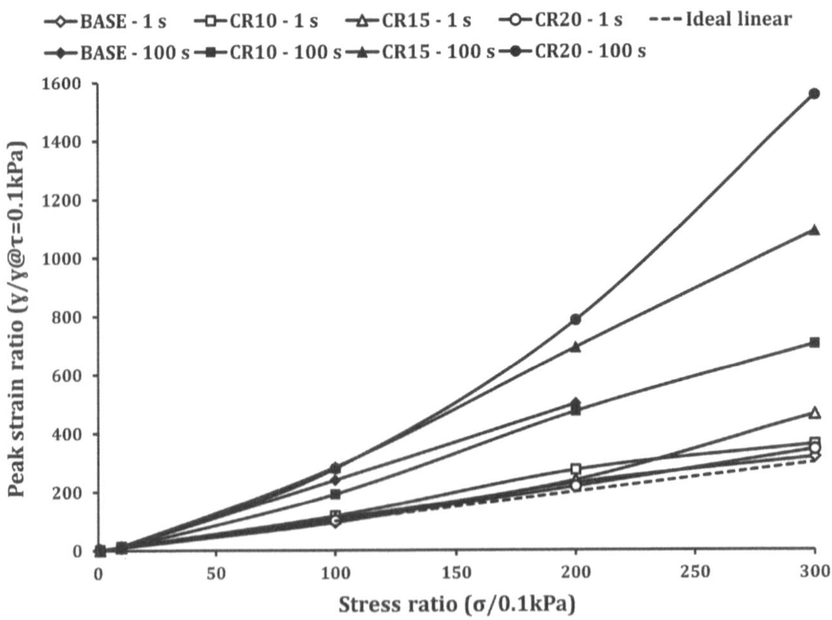

Figure 7.16 Peak strain ratio versus stress ratio for both creep loading time of 1 s and 100 s and different dosages of crumb rubber (CR)–modified bitumen.

Recalling Eq. (3.65) for creep compliance of GVFM, it is possible to use this equation for each stress level individually to determine constant parameters of A_i and β_i. To consider the effect of stress level on the amount of strain in Eq. (3.65), a normalized multiplier for stress can be added to this equation as follows:

$$\gamma(t) = \sum_{i=1}^{n} A_i \, t^{\beta_i} \sigma^{m_i} \tag{7.33}$$

in which t is normalized, $t = t/(t = 1 \text{ s})$, and σ is normalized stress, $\sigma = \sigma/(\sigma = \text{minimum stress level})$. Considering Eq. (7.33), it is obvious that for an ideal linear viscoelastic behavior, the power of normalized stress, m, should be 1 and strain at different stress levels can be calculated by multiplying strain at a specified stress level to the amount of stress level according to the main assumption of linear viscoelastic theorem (Ottosen & Ristinmaa, 2005). Hajikarimi et al. (2018a,b) implemented a binomial expression for the GVFM, as shown in Eq. (7.34), to evaluate the parameter of m as the nonlinearity index.

$$\gamma(t) = A_1 t^{\beta_1} \sigma + A_2 t^{\beta_2} \sigma^{m} \tag{7.34}$$

The power of σ in the first expression of Eq. (7.34) was considered to be 1 to reflect the linear part of viscoelastic strain and all nonlinear properties were reflected in the power of σ in the second expression as the parameter of m. To calculate the parameter of m, a curve-fitting procedure can be used to fit Eq. (7.34) on experimental data. Fig. 7.17 illustrates a 3D plane fitted on strain data of the original bitumen. With regard to Eq. (7.34), it is evident that the higher amount of m shows a higher amount of nonlinearity for bituminous materials.

It is possible to investigate the effect of modification on recovery characteristics of bitumen; the proportion of permanent strain to peak strain of creep and recovery test should be used. After removing the creep load, part of the strain will be removed, which is known as recovered strain, and the remaining part of strain is known as permanent strain. A simple equation can be represented as follows for the abovementioned descriptions:

$$\gamma(t, \sigma) = \gamma_r(t, \sigma) + \gamma_p(t, \sigma) \tag{7.35}$$

where γ_r is recovered strain and γ_p is permanent strain. The greater amount of recovered strain shows the ability of bitumen to resist against

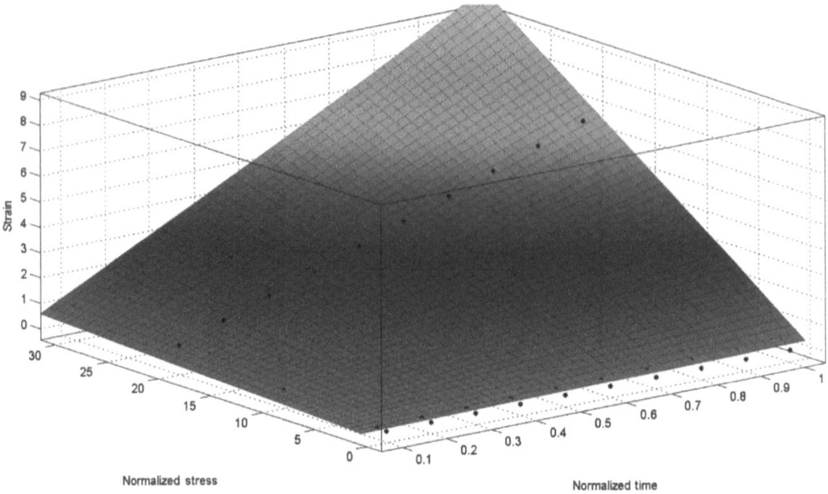

Figure 7.17 Illustration of a three-dimensional plane fitted on strain data of the original bitumen to investigate nonlinear viscoelastic behavior. *From Hajikarimi, P., Nejad, F. M., Khodaii, A., & Fini, E. H. (2018). Introducing a stress-dependent fractional nonlinear viscoelastic model for modified asphalt binders. Construction and Building Materials, 183, 102–113.*

permanent deformation. By increasing the stress level, the ratio of permanent strain to total strain decreases. To model the recovery behavior of original and modified bitumen by using the generalized fractional viscoelastic model, it is possible to add some terms to Eq. (7.34) that work after the time of removing creep load as:

$$\gamma(t) = A_1 t^{\beta_1}\sigma + A_2 t^{\beta_2}\sigma^m - A_3(t-t_1)^{\beta_3}\sigma - A_4(t-t_1)^{\beta_4}\sigma^n \qquad (7.36)$$

in which t_1 is the time of load removal. As can be seen in Eq. (7.36), parameters of A_3, A_4, β_3, β_4, and n work after load removal to determine captured strain in the recovery zone.

7.3.5 Repeated creep and recovery

The RCR test was developed by Bahia et al. (2001) during the NCHRP 9-10 project to evaluate bitumen's resistance to permanent deformation. The RCR is a cyclic test proposed as a method of separating the dissipated energy and evaluating the resistance to accumulation of permanent strain for bitumens as an index of rutting susceptibility. The NCHRP 9-10 project proposed a shear stress in the range of 30–300 Pa for 100 cycles with 1 s loading time followed by 9 s unloading time.

Bahia, Hanson et al. (2001) introduced a new parameter Gv to overcome weakness of superpave specification parameter, $G^*/\sin\delta$, and characterize an appropriate rutting resistance indicator of bitumens. This parameter was entitled as the viscous component of creep stiffness which can be derived by using the Burgers model (see Section 3.2.3) based on the RCR test results. The total shear strain in the Burgers model is expressed as follows (Wang et al., 2016):

$$\gamma(t) = \gamma_E + \gamma_{DE} + \gamma_V = \frac{\tau_0}{G_0} + \frac{\tau_0}{G_1}\left(1 - \exp\left(-\frac{G_1}{\eta_1}t\right)\right) + \frac{\tau_0}{\eta_0}t \quad (7.37)$$

where $\gamma(t)$ is total shear strain; $\gamma_E(t)$ is elastic shear strain; $\gamma_{DE}(t)$ is delayed elastic shear strain; $\gamma_V(t)$ is viscous shear strain; τ_0 is constant stress; G_0 is instantaneous elastic modulus of the Maxwell model; G_1 is spring constant of the Kelvin–Voigt model; η_0 is viscosity coefficient of the Maxwell model; η_1 is viscosity coefficient of the Kelvin–Voigt model; and t is time.

Knowing that creep compliance $J(t) = \gamma(t)/\tau_0$, we have:

$$J(t) = J_E + J_{DE}(t) + J_V(t) \quad (7.38)$$

in which $J(t)$ is creep compliance; $J_E(t)$ is elastic creep compliance; $J_{DE}(t)$ is delayed elastic creep compliance; and $J_V(t)$ is viscous creep compliance.

The viscous component of the creep stiffness, G_V was defined as the inverse of viscous creep compliance as follows:

$$G_v = \frac{1}{J_v} = \frac{\eta_0}{t} \quad (7.39)$$

The dissipated energy could be determined in terms of the four Burgers parameters (Johansson & Isacsson, 1998). The dissipated energy is inversely related to the viscosity coefficient of the Maxwell's dashpot. Thus the dissipated energy decreases by increasing the viscous component of creep stiffness. Accordingly, increasing Gv causes decreased dissipated energy and antirutting potential should reduce.

Bahia, Zhai et al. (2001) suggested using cycles 50 and 51 to fit the response of the RCR test utilizing the Burgers model. It is necessary to mention that determining the four Burgers parameters needs to consider creep and recovery simultaneously. Therefore it is necessary to derive the equation of recovery for the Burgers model. Fig. 7.18 schematically shows strain versus time for one cycle of the RCR test. If the stress τ_0 is removed at time t_1, the recovery behavior of the Burgers model can be determined from Eq. (7.37) and the superposition

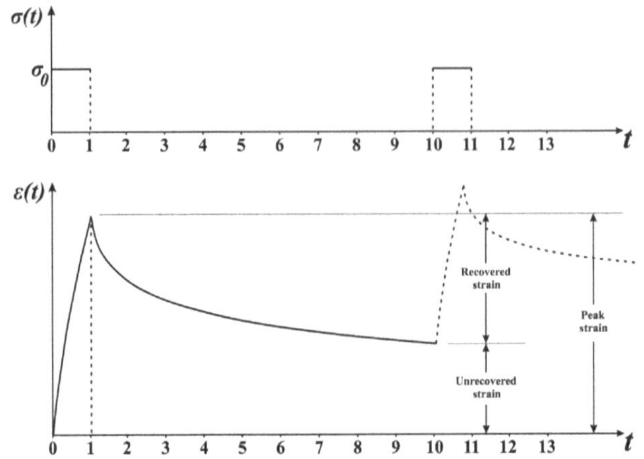

Figure 7.18 Strain versus time for one cycle of the repeated creep and recovery test.

principle by considering that at $t = t_1$ a constant stress $\tau = -\tau_0$ is added. The recovery strain $\varepsilon(t)$, for $t > t_1$ is the sum of these two independent actions (Findley, Lai, & Onaran, 1976):

$$\begin{aligned}\varepsilon(t) = {} & \frac{\tau_0}{G_0} + \frac{\tau_0}{\eta_0}t + \frac{\tau_0}{G_1}\left(1 - \exp\left(-\frac{G_1}{\eta_1}t\right)\right) \\ & - \left[\frac{\tau_0}{G_0} + \frac{\tau_0}{\eta_0}(t - t_1) + \frac{\tau_0}{G_1}\left(1 - \exp\left(-\frac{G_1}{\eta_1}(t - t_1)\right)\right)\right] \\ = {} & \frac{\tau_0}{\eta_0}t_1 + \frac{\tau_0}{G_1}\left(\exp\left(\frac{G_1}{\eta_1}t_1\right) - 1\right)\exp\left(-\frac{G_1}{\eta_1}t\right)\end{aligned} \tag{7.40}$$

The fitting procedure for determining the four Burgers parameters should be performed on the curve of strain versus time. Based on Eq. (7.39), to determine the viscous component of creep stiffness, it is just required to calculate η_0 as one of the four Burgers model.

7.3.6 Multiple stress creep and recovery

Since single stress level in the RCR test cannot thoroughly evaluate the stress dependency of modified binders (D'Angelo et al., 2007; Golalipour, 2011), this test was revised as the MSCR test (ASTM D7405-08a) to properly characterize bitumens properties in terms of their rutting resistance (Tabatabaee & Tabatabaee, 2010). The MSCR test method includes 1 s creep loading followed by 9 s recovery under two different stress levels of 0.1 and 3.2 kPa at 10 cycles for each stress level (10 cycles with 0.1 kPa loading and 10 cycles with 3.2 kPa loading). It is evident that this test must be performed at high temperatures to investigate rutting susceptibility of original and modified bitumens. As an example, the results of MSCR test at 50°C have been shown in Fig. 7.19 for original and nanosilica-modified bitumens. During the loading phase, the instantaneous elastic strain of bitumen was measured as well as the viscoelastic strain of bitumen during the unloading stage to represent the total creep strain accumulated. The instantaneous elastic strain of bitumen disappeared after unloading while the delayed elastic strain recovered gradually. This is when the nonrecoverable strain is left after unloading; this portion of strain is related to permanent strain.

The average nonrecoverable creep compliance (J_{nr}) and the average percentage of recoverable strain (R%) are two primary parameters of the MSCR test to evaluate the resistance of bitumen against rutting (J. D'Angelo, 2009; J. A. D'Angelo, 2009). At each stress level, the average nonrecoverable creep compliance is defined by dividing the average nonrecoverable strain for the 10 creep and recovery cycles to the applied

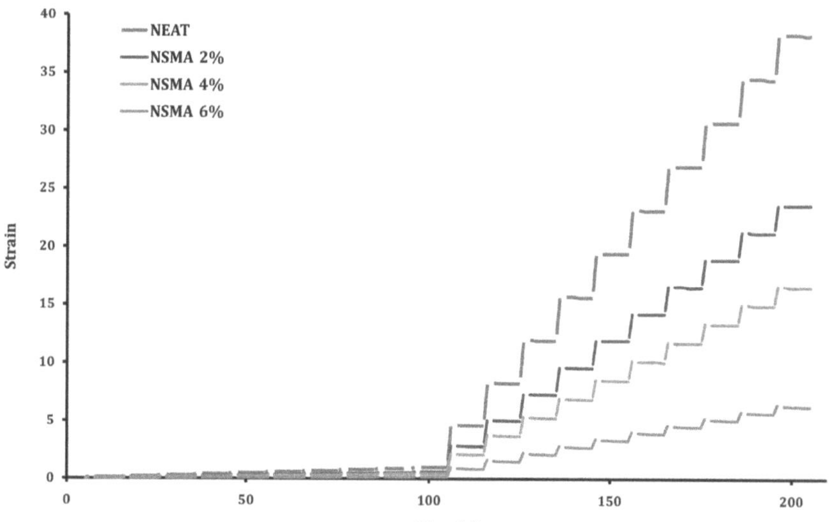

Figure 7.19 Results of multiple stress creep and recovery test at 50°C for original and nanosilica-modified bitumens.

stress for those cycles. The calculation procedure for J_{nr} at the stress level of σ [$J_{nr}(\sigma)$] is as follows (ASTM D7405-20, 2020):

$$Jnr(\sigma,\ N) = \frac{\xi_{10}}{\sigma} \tag{7.41}$$

$$Jnr(\sigma) = \frac{\sum_{N=1}^{10} Jnr(\sigma,\ N)}{10} \tag{7.42}$$

where

$$\xi_{10} = \xi_r - \xi_0$$

and ξ_r is strain value at the end of the recovery portion (i.e., after 10.0 s) of each cycle strain, and ξ_0 is initial strain value at the beginning of the creep portion of each cycle.

The J_{nr} parameter was introduced as an index of the bitumen's contribution to mixture rutting performance. Lower J_{nr} denotes better resistance to rutting. Also, following equations show the calculation method for %R at the stress level of σ.

$$\xi_r(\sigma,\ N) = \frac{(\xi_1 - \xi_{10}) \times 100}{\xi_1} \tag{7.43}$$

$$R(\sigma) = \frac{\sum_{N=1}^{10} \xi_r(\sigma,\ N)}{10} \tag{7.44}$$

where

$$\xi_1 = \xi_c - \xi_0$$

and ξ_c is strain value at the end of the creep portion of each cycle, and ξ_0 is initial strain value at the beginning of the creep portion of each cycle.

The average percent of recoverable strain determines the elastic behavior of bitumens under loading. Elasticity is an important property that measures the ability of materials to recover after deformation. A bitumen with higher percent of recoverable strain has lower tendency to permanent deformation.

7.3.7 Linear amplitude sweep test

There are some well-known experimental methods for evaluating fatigue resistance of bituminous materials. The contribution of bitumen to asphalt concrete's fatigue resistance can be accurately determined using a multiple accelerated testing procedure (Hintz & Bahia, 2013b). The LAS test is a new test for investigating fatigue behavior of original and modified bitumens (Hintz & Bahia, 2013a). Based on viscoelastic continuum damage (VECD) mechanics, this test can estimate fatigue life cycle of bitumen as a function of strain in the pavement structure. During the procedure of running the LAS test, using cyclic loading with an increasing regime of load amplitude, damage can be accelerated and sufficient data can be collected for analysis in less than 30 min (Notani et al., 2019).

The LAS test includes two parts. The first part is a frequency sweep test (see Section 7.3.1) for acquiring undamaged rheological characteristics of bitumens immediately after applying a low-frequency load and the second part is determining bitumen's fatigue characteristics implementing different strain levels during 30 min.

The frequency sweep data is used to calculate the damage element of alpha which is a function of slope G' (storage modules) and ω, while G' can be determined from Eq. (7.45):

$$G'(\omega) = |G^*|(\omega) \times \cos\delta(\omega) \tag{7.45}$$

The alpha parameter was calculated from the inverse of the slope of the plot of log G' versus log ω as follows:

$$\log G'(\omega) = m\log(\omega) + b \tag{7.46}$$

$$\alpha = \frac{1}{m} \tag{7.47}$$

The susceptibility of bitumen against fatigue distresses can be related to accumulation damage. To obtain accumulation damage of bitumens, the following equation was utilized:

$$D(t) = \sum_{i=1}^{n} \left[\prod \gamma_0{}^2 (C_{i-1} - C_i)\right]^{\frac{\alpha}{1+\alpha}} (t_i - t_{i-1})^{\frac{1}{1+\alpha}} \tag{7.48}$$

where $\gamma_0{}^2$ is applied strain at a given data point.

$$C_i = \frac{|G^*| \times \sin\delta(t)}{|G^*|_0 \times \sin\delta_{\text{initial}}}$$

where $|G^*|$ is complex shear modulus, MPa; α is damage parameter as defined in Eq. (7.46) and Eq. (7.47); and t is time in seconds.

For each data point at a given time, $G^*.\sin\delta(t)$ and $D(t)$can be determined. It is assumed that C is equal to 1 at $D(0)$ and $D(0) = 0$. The following equation shows the relationship between C and $D(t)$. C_0, C_1, and C_2 are three required parameters for this equation which can be determined by using a simple fitting procedure of power law functions:

$$C = C_0 - C_1(D)^{C_2} \tag{7.49}$$

The power relation of Eq. (7.49) can be linearized as:

$$\log(C_0 - C) = \log(C_1) + C_2 \times \log(D) \tag{7.50}$$

In which C_1 is the intercept and C_2 is the slope of plot of $\log(C_0 - C)$ versus $\log(D)$. The data points with accumulation damage of less than 10 are ignored during the fitting procedure. Fig. 7.20 schematically shows the parameter C versus damage intensity.

The value of $D(t)$ at failure, D_f, is defined as $D(t)$ in which corresponding fatigue parameter ($|G^*|\sin\delta$) reduces 35% rather than initial $|G^*|$ $\sin\delta$. The D_f is calculated by using the following equation:

$$D_f = 0.35\left(\frac{C_0}{C_1}\right)^{(1/C_2)} \tag{7.51}$$

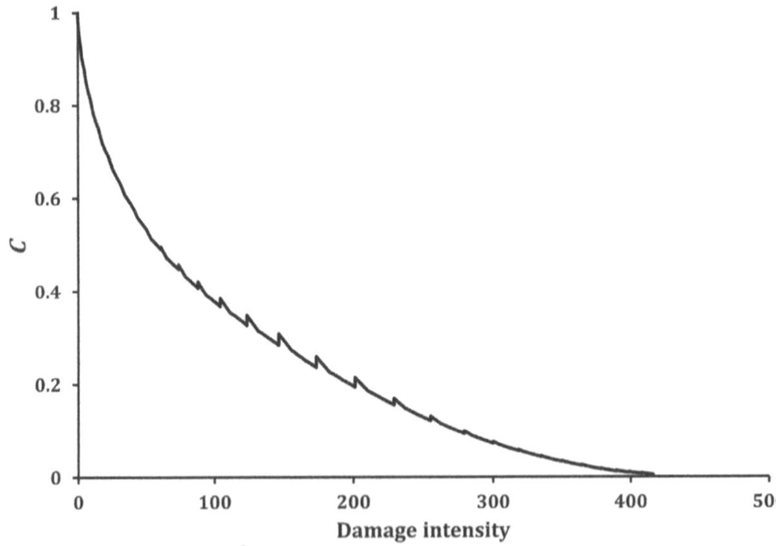

Figure 7.20 Schematic plot of C versus damage intensity for linear amplitude sweep test.

To predict the bitumen fatigue performance parameter of N_f, two bitumen parameters of A and B are required. These parameters (A and B) of the model can be calculated as follows:

$$A = \frac{f\left(D_f\right)^k}{k(\pi I_D C_1 C_0)^\alpha} \tag{7.52}$$

in which f is loading frequency (10 Hz) and $k = 1 + (1 - C_2)\alpha$ and

$$B = 2\alpha \tag{7.53}$$

Finally, the bitumen fatigue parameter (N_f) can be calculated by using Eq. (7.54):

$$N_f = A\left(\gamma_{max}\right)^{-B} \tag{7.54}$$

where γ_{max} is the maximum expected bitumen strain for a specific pavement structure.

When a circumferential crack initiates at the outside boundary of the bitumen sample and propagates inward during the fracture process, the sample radius (r) reduces by increasing the number of loading cycles. It is understandable that cracking leads to apparent modulus reduction. Crack length can be predicted by using changes in torsional stiffness, k, which can be determined as below:

$$k = \frac{T}{\varphi} \tag{7.55}$$

where k is torsional stiffness, T is torque, and φ is deflection angle. Fig. 7.21 shows an example curve showing torque versus number of cycles. The effective radius for each given data point can be calculated based on changing torsional stiffness. The reduced effective radius can be represented by the following relationship:

$$r_N{}^4 = r_i{}^4 \frac{k_n}{k_i} \tag{7.56}$$

where r_N is the effective sample radius at cycle N (mm); r_i is initial sample radius (4 mm); k_n is torsional stiffness at cycle N; and k_i is initial torsional stiffness.

The effective crack length can be determined based on the sample radius of each cycle using the following equation:

$$a = r_i - r_N \tag{7.57}$$

in which a is effective crack length.

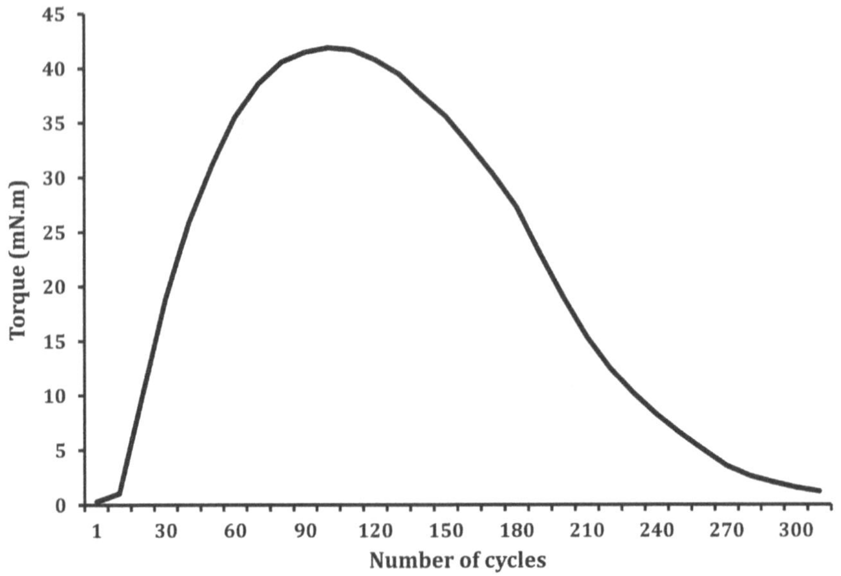

Figure 7.21 An example for illustrating torque versus number of cycles in linear amplitude sweep test.

Hintz and Bahia (2013b) also introduced a relation for determining energy release rate as follows:

$$G_f = \frac{|G^*|_0 \cdot \gamma_i^2 . h}{2 . r_i^2 (r_i - a)} (r_i - a + c . (1 - e^{-a/c}))^3 . (1 - e^{-a/c}) \tag{7.58}$$

In which $|G^*|_0$: the undamaged complex modulus; h is the sample height; r_i is the initial sample radius (4 mm); a is the crack length; and c is numerical factor equal to 0.1 for bitumens tested implementing 8 mm parallel plate geometry.

7.3.8 Healing characteristics

To investigate the healing properties of neat and modified asphalt binders in macroscale, a DSR test setup was introduced by Bhasin et al. (2008) as a two-piece healing (TPH) test. In this method (as can be seen in Fig. 7.22), two pieces of bitumen with diameter of 25 mm and height of 2.5 mm were attached to the spindle and the bottom plate of the DSR. These two samples are moved toward each other and the gap is set to 5 mm. By closing two pieces together until the defined gap value,

Figure 7.22 Setup of two-piece healing test.

complex modulus (G^*) was measured at 0, 2, 4, 6, 8, 10, 15, 20, 25, 30, 40, 50, and 60 min while a strain of 0.001% was applied. All tests were performed at 25°C and normal force was set on zero. This protocol was also done for an integrated (one-piece) sample with the same diameter and height of 5 mm. The average G^* at each time after three replicates was calculated for each sample and the proportion of G^* (two-piece sample) and G^* of one-piece sample can represent the percentage of healing.

As an example, Fig. 7.22 depicts percentage of healing [proportion of G^* (two-piece sample) and G^* of one-piece sample] for neat and PPA-modified asphalt binders versus time during the TPH test. Implementing results of TPH test, it is possible to determine parameters of healing function (Bhasin et al., 2008) including R_0, p, q, and r as follows:

$$R_h(t) = R_0 + p(1 - \exp(-qt^r)) \tag{7.59}$$

In the above equation, $R_h(t)$ can be used to show the increase in mechanical property of the wetted crack interface as a time dependent function. It is possible to use Eq. (7.59) for any mechanical property. The parameter R_0 represents the effect of instantaneous healing of wetted surfaces. This instantaneous healing relates to the interfacial cohesion driven by surface free energy. The second part of Eq. (7.59) reflects the kinetics of a phase transformation because of the migration and randomization of molecules across the interface. Actually, the second part of this equation relates to long-term healing properties. It is necessary to mention that

p, q, and r are time-independent parameters which can be calculated implementing the principle of multiple nonlinear regression. These parameters are dependent to intrinsic material properties, such as molecular weight, and activation energy for diffusion, and also extrinsic parameters, such as pressure and temperature.

7.4 Rotational viscometer

The rotational viscosity is used to evaluate high-temperature workability of bitumens. A rotational coaxial cylinder viscometer known as "Brookfield viscosity" is widely used for this purpose to measure viscosity according to AASHTO TP48. The viscosity of bitumen is measured at high temperatures to ensure that the bitumen is fluid enough during the pumping and mixing procedure. Therefore the rotational viscosity should be measured on unaged bitumen to accurately simulate the aging condition of bitumen.

Rotational viscosity is determined by measuring the torque required to maintain a constant rotational speed of a cylindrical spindle while submerged in a sample at a constant temperature. The torque is directly related to the viscosity of the bitumen sample. Fig. 7.23 schematically shows a rotational viscometer.

The rotational viscosity of bitumen should be measured at different high temperatures to sketch the plot of viscosity versus temperature as shown in Fig. 7.24 for an original and ethylene bis-stearamide (EBS)-modified bitumen. By using such a plot, it will be possible to specify proper mixing and compaction temperature to meet the desired viscosity range for these conditions, which is 170 ± 20 cSt for mixing temperature and 280 ± 30 cSt for compaction temperature. For example, considering

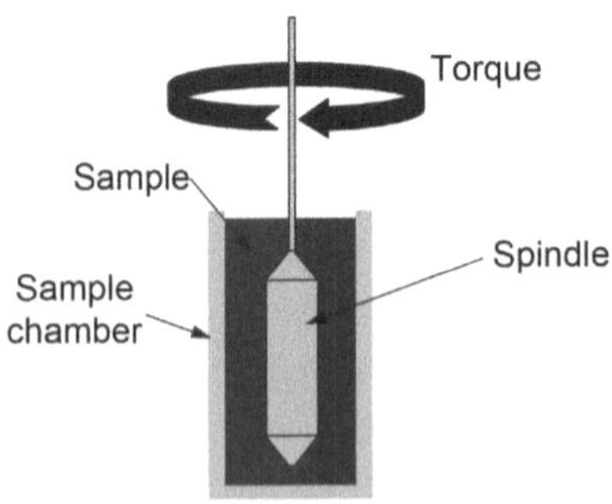

Figure 7.23 Schematic representation of a rotational viscometer.

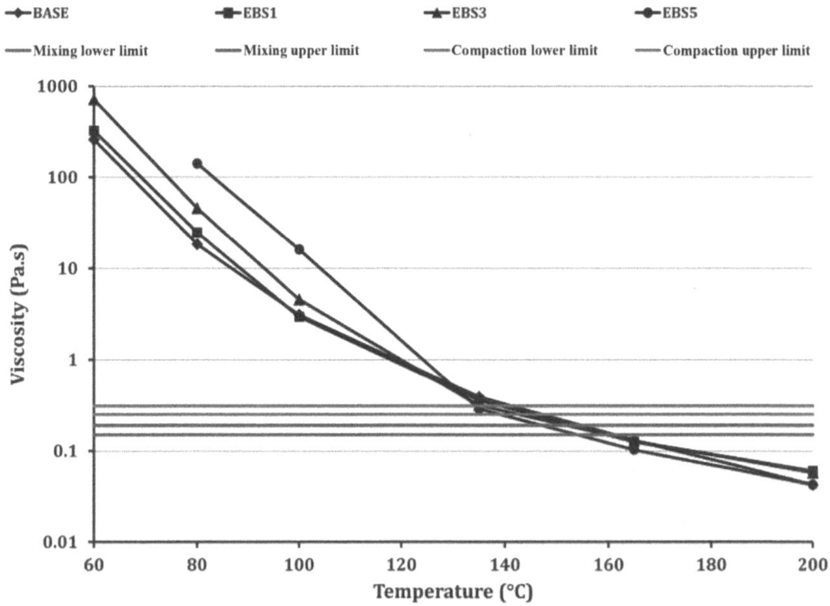

Figure 7.24 Viscosity versus temperature for an original and EBS-modified bitumen.

Fig. 7.24, it is evident that adding EBS to the original bitumen decreases mixing and compaction temperature of asphalt mixture which is known as warm mix asphalt additive.

References

Aflaki, S., & Hajikarimi, P. (2012). Implementing viscoelastic rheological methods to evaluate low temperature performance of modified asphalt binders. *Construction and Building Materials, 36*. Available from https://doi.org/10.1016/j.conbuildmat.2012.04.076.

Anderson, D. A., Christensen, D. W., Bahia, H. U., Dongre, R., Sharma, M. G., Antle, C. E., & Button, J. (1994). *Binder characterization and evaluation, volume 3: Physical characterization. Strategic Highway Research Program* (Report No. SHRP-A-369). National Research Council.

Anderson, D. A., Christensen, D. W., Dongre, R., Sharma, M. G., Runt, J., & Jordhal, P. (1990). *Asphalt behavior at low service temperatures. Final Report.* National Technical Reports Library, U.S. Department of Commerce.

Anderson, D. A., Le Hir, Y. M., Planche, J.-P., Martin, D., & Shenoy, A. (2002). Zero shear viscosity of asphalt binders. *Transportation Research Record, 1810*(1), 54–62.

ASTM D6648-08 (2001). *Standard test method for determining the flexural creep stiffness of asphalt binder using the bending beam rheometer (BBR).* West Conshohocken, PA: ASTM International.

ASTM D175M-09 (2014). *Standard test method for effects of heat and air on asphaltic materials (thin-film oven test).* West Conshohocken, PA: ASTM International.

ASTM D6521-19a (2019). *Standard practice for accelerated aging of asphalt binder using a pressurized aging vessel (PAV).* West Conshohocken, PA: ASTM International.

ASTM D7405-20. (2020). Standard test method for multiple stress creep and recovery (MSCR) of asphalt binder using a dynamic shear rheometer. West Conshohocken, PA: ASTM International.

Bahia, H. U., Anderson, D. A., & Christensen, D. W. (1992). The bending beam rheometer; a simple device for measuring low-temperature rheology of asphalt binders (with discussion). Journal of the Association of Asphalt Paving Technologists, *61*, 117–153.

Bahia, H. U., Hanson, D. I., Zeng, M., Zhai, H., Khatri, M. A., & Anderson, R. M. (2001). *Characterization of modified asphalt binders in superpave mix design* (Issue Project 9-10 FY'96; NCHRP Report No. 459). Washington, DC: National Academy Press.

Bahia, H. U., Zhai, H., Zeng, M., Hu, Y., & Turner, P. (2001). Development of binder specification parameters based on characterization of damage behavior (with discussion). *Journal of the Association of Asphalt Paving Technologists*, *70*, 442–470.

Bhasin, A., Little, D. N., Bommavaram, R., & Vasconcelos, K. (2008). A framework to quantify the effect of healing in bituminous materials using material properties. *Road Materials and Pavement Design*, *9*(sup1), 219–242. Available from https://doi.org/10.1080/14680629.2008.9690167.

D'Angelo, J. (2009). Current status of superpave binder specification. *Road Materials and Pavement Design*, *10*(sup1), 13–24. Available from https://doi.org/10.1080/14680629.2009.9690233.

D'Angelo, J. A. (2009). The relationship of the MSCR test to rutting. *Road Materials and Pavement Design*, *10*(sup1), 61–80. Available from https://doi.org/10.1080/14680629.2009.9690236.

D'Angelo, J., Kluttz, R., Dongre, R. N., Stephens, K., & Zanzotto, L. (2007). Revision of the superpave high temperature binder specification: The multiple stress creep recovery test (with discussion). *Journal of the Association of Asphalt Paving Technologists*, *76*, 123–162.

Eschbach, A. R. (1993). *Dynamic shear rheometer and method*. Google Patents.

Findley, W. N., Lai, J. S., & Onaran, K. (1976). *Creep and relaxation of nonlinear materials*. Amsterdam: Courier Corporation.

Fini, E. H., Hajikarimi, P., Rahi, M., & Nejad, F. M. (2016). Physiochemical, rheological, and oxidative aging characteristics of asphalt binder in the presence of mesoporous silica nanoparticles. *Journal of Materials in Civil Engineering*, *28*(2), Article 04015133. Available from https://doi.org/10.1061/(ASCE)MT.1943-5533.0001423.

Golalipour, A. (2011). *Modification of multiple stress creep and recovery test procedure and usage in specification* (Master's thesis). Madison, WI: University of Wisconsin.

Gooding, D., Pinch, T., & Schaffer, S. (1989). *The uses of experiment: Studies in the natural sciences*. Cambridge University Press.

Guericke, R., & Hoeppel, H.E. (2000). *Results of ARBIT test programme 1998/99 involving 36 unmodified and modified bituminous binders on the German market*. Proceedings of the papers submitted for review at 2nd EURASPHALT and EUROBITUME Congress, September 20–22, 2000, Barcelona, Spain. Book 1-Session 1.

Hajikarimi, P., Aflaki, S., & Hoseini, A. S. (2013). Implementing fractional viscoelastic model to evaluate low temperature characteristics of crumb rubber and gilsonite modified asphalt binders. *Construction and Building Materials*, *49*, 682–687. Available from https://doi.org/10.1016/j.conbuildmat.2013.09.001.

Hajikarimi, P., Moghadas Nejad, F., Khodaii, A., & Fini, E. H. (2018a). Introducing a stress-dependent fractional nonlinear viscoelastic model for modified asphalt binders. *Construction and Building Materials*, *183*, 102–113. Available from https://doi.org/10.1016/j.conbuildmat.2018.06.166.

Hajikarimi, P., Nejad, F. M., & Aghdam, M. M. (2018b). Implementing general power law to interconvert linear viscoelastic functions of modified asphalt binders. *Journal of Transportation Engineering Part B: Pavements*, *144*(2). Available from https://doi.org/10.1061/JPEODX.0000038.

Hajikarimi, P., Rahi, M., & Moghadas Nejad, F. (2015). Comparing different rutting specification parameters using high temperature characteristics of rubber-modified asphalt binders. *Road Materials and Pavement Design, 16*(4), 1–16. Available from https://doi.org/10.1080/14680629.2015.1063533.

Hintz, C., & Bahia, H. (2013a). Simplification of linear amplitude sweep test and specification parameter. *Transportation Research Record, 2370*(1), 10–16.

Hintz, C., & Bahia, H. (2013b). Understanding mechanisms leading to asphalt binder fatigue in the dynamic shear rheometer. *Road Materials and Pavement Design, 14*(sup2), 231–251. Available from https://doi.org/10.1080/14680629.2013.818818.

Hoare, T. R., & Hesp, S. A. M. (2000). Low-temperature fracture testing of asphalt binders: Regular and modified systems. *Transportation Research Record, 1728*(1), 36–42.

Johansson, L. S., & Isacsson, U. (1998). Effect of filler on low temperature physical hardening of bitumen. *Construction and Building Materials, 12*(8), 463–470. Available from https://doi.org/10.1016/S0950-0618(98)00028-2.

Kamiar, M. (2008). *Brilliant Biruni: A life story of Abu Rayhan Mohammad Ibn Ahmad.* Scarecrow Press.

Kim, R. Y. (2009). *Modeling of asphalt concrete.* Mc-Graw Hill.

Kim, Y. R., Lee, H.-J., & Little, D. N. (1997). Fatigue characterization of asphalt concrete using viscoelasticity and continuum damage theory (with discussion). *Journal of the Association of Asphalt Paving Technologists, 66*, 520–569.

Li, X., & Marasteanu, M. (2004). Evaluation of the low temperature fracture resistance of asphalt mixtures using the semi circular bend test. *Association of Asphalt Paving Technologists, 73*, 401–426.

Liu, S., Cao, W., Shang, S., Qi, H., & Fang, J. (2010). Analysis and application of relationships between low-temperature rheological performance parameters of asphalt binders. *Construction and Building Materials, 24*(4), 471–478. Available from https://doi.org/10.1016/j.conbuildmat.2009.10.015.

Marasteanu, M., Zofka, A., Turos, M., Li, X., Velasquez, R., Li, X., Buttlar, W., Paulino, G., Braham, A., & Dave, E. (2007). *National Pooled Fund Study 776—Investigation of low temperature cracking in asphalt pavements.* St. Paul, MN: Minnesota Department of Transportation.

Mobasher, B., Mamlouk, M. S., & Lin, H.-M. (1997). Evaluation of crack propagation properties of asphalt mixtures. *Journal of Transportation Engineering, 123*(5), 405–413.

Molenaar, A. A. A., Scarpas, A., Liu, X., & Erkens, S. (2002). Semi-circular bending test; simple but useful? *Journal of the Association of Asphalt Paving Technologists, 71*, 794–815.

Notani, M. A., Moghadas Nejad, F., Khodaii, A., & Hajikarimi, P. (2019). Evaluating fatigue resistance of toner-modified asphalt binders using the linear amplitude sweep test. *Road Materials and Pavement Design, 20*(8), 1927–1940. Available from https://doi.org/10.1080/14680629.2018.1474792.

Ottosen, N. S., & Ristinmaa, M. (2005). *The mechanics of constitutive modeling.* Elsevier.

Shapere, D. (1982). The concept of observation in science and philosophy. *Philosophy of Science, 49*(4), 485–525.

Shaw, M. T., & MacKnight, W. J. (2018). *Introduction to polymer viscoelasticity.* John Wiley & Sons.

Sybilski, D. (1994). Relationship between absolute viscosity of polymer-modified bitumens and rutting resistance of pavement. *Materials and Structures, 27*(2), 110–120.

Sybilski, D. (1996). Zero-shear viscosity of bituminous binder and its relation to bituminous mixture's rutting resistance. *Transportation Research Record, 1535*(1), 15–21.

Tabatabaee, N., & Tabatabaee, H. A. (2010). Multiple stress creep and recovery and time sweep fatigue tests: Crumb rubber modified binder and mixture performance. *Transportation Research Record, 2180*(1), 67–74.

Tayebali, A. A., Rowe, G. M., & Sousa, J. B. (1992). Fatigue response of asphalt-aggregate mixtures (with discussion). *Journal of the Association of Asphalt Paving Technologists*, *61*, 333–360.

Timoshenko, S., & Young, D. H. (1962). *Elements of strength of materials*. Princeton, NJ: Van Nostrand.

Wagnoner, M. P., Buttlar, W. G., & Paulino, G. H. (2005). Disk-shaped compact tension test for asphalt concrete fracture. *Experimental Mechanics*, *45*(3), 270–277. Available from https://doi.org/10.1007/BF02427951.

Wagoner, M. P., Buttlar, W. G., & Paulino, G. H. (2005). Development of a single-edge notched beam test for asphalt concrete mixtures. *Journal of Testing and Evaluation*, *33* (6), 452–460.

Wagoner, M. P., Buttlar, W. G., Paulino, G. H., & Blankenship, P. (2006). Laboratory testing suite for characterization of asphalt concrete mixtures obtained from field cores (with discussion). *Journal of the Association of Asphalt Paving Technologists*, *75*, 815–851.

Wang, C., Zhang, H., Castorena, C., Zhang, J., & Kim, Y. R. (2016). Identifying fatigue failure in asphalt binder time sweep tests. *Construction and Building Materials*, *121*, 535–546. Available from https://doi.org/10.1016/j.conbuildmat.2016.06.020.

Yildirim, Y., Solaimanian, M., & Kennedy, T.W. (2000). Mixing and compaction temperatures for hot mix asphalt concrete (Research Report No. 1250-5). Austin, TX: Center for Transportation Research.

Youngguk, A. (2003). *A comprehensive study of crack growth in asphalt concrete using fracture mechanics* (Doctoral dissertation, Civil Engineering). Raleigh, NC: North Caroline State University.

CHAPTER 8

Incremental form of viscoelastic relations

8.1 Introduction

As mentioned in Chapter 1, Introduction to Viscoelasticity, stresses, strains, and displacements are three types of quantities that needed to be determined in the mechanics of continuum media. In the elastic domain, there are closed-form relations to connect these parameters which can be found in several textbooks (Barber, 2002; Sadd, 2009; Soutas-Little, 2012). However, viscoelastic materials are more complicated as their mechanical behavior depends on time, temperature, and loading rate. With regard to Boltzmann's superposition principle (see Chapter 2: Constitutive Viscoelastic Functions), it is possible to discretize the whole time to several time steps to use some simplifying assumptions for determining stresses and strains. In Chapter 4, Correspondence Principle of Viscoelasticity, the correspondence principle was introduced to implement closed-form elastic solutions for solving different types of problems in the viscoelastic domain.

There are several problems with particular boundaries and loading conditions for which it is not possible to develop a closed-form solution even in the elastic domain. In such cases, it is required to use numerical methods of which the finite element method (FEM) is the most widely used one for this purpose. The FEM subdivides the considered medium, which is usually large, into smaller, simpler, and known parts which are called finite elements (FEs). For this purpose, a specific space discretization must be implemented by constructing a mesh on the considered domain having a finite number of points. This method consequently resulted in a system of algebraic equations for approximating the unknown functions over the domain and solving

Applications of Viscoelasticity
DOI: https://doi.org/10.1016/B978-0-12-821210-3.00004-8

partial differential equations that do not have an explicit solution (Zienkiewicz & Morice, 1971). Traditionally, the FEM was used to solve problems in the field of structural analysis, heat transfer, fluid flow, and mass transport (Reddy, 1993). Other numerical methods were also inspired by the FEM, including extended finite element method (XFEM) (Chessa et al., 2002; Mohammadi, 2008), boundary element method (Banerjee & Butterfield, 1981), and mesh-free method (Liu & Gu, 2005).

To implement the FEM for solving problems in the viscoelastic domain, it is required to use incremental relations (considering discretizing the whole time to a number of time steps) to determine stresses and strains. Therefore the main focus of this chapter is to develop the incremental form of viscoelastic relations by considering both integer-order and fractional-order constitutive viscoelastic equations, which can cover a wide range of problems. It is inevitable to present a simple FEM framework to show how the incremental form of viscoelastic relations can be used to solve simple and practical problems. Then, different types of examples are used in this chapter to show the ability of this method as well as providing a MATLAB® code that can be used as a function in the FEM framework.

8.2 Incremental form of viscoelastic relations

In this section, a detailed and step-by-step procedure is presented to develop an incremental form of viscoelastic relations for both integer-order and fractional-order viscoelastic models which are thoroughly introduced in Chapter 3, Mechanical Models of Viscoelasticity. Each type of viscoelastic model can be implemented for a particular case.

8.2.1 Integer-order viscoelastic models

In the uniaxial condition, the stress–strain relation can be described based on Boltzmann's superposition integration as follows:

$$\sigma(t) = E(t)\,\varepsilon(0) + \int_{-\infty}^{t} E(t-\xi)\frac{d\varepsilon(\xi)}{d\xi}\,d\xi \tag{8.1}$$

in which $\sigma(t)$ and $\varepsilon(t)$ are stress and strain, respectively and $E(t)$ is relaxation modulus which can be determined in the form of Prony series as an integer-order viscoelastic model, or power law as a fractional-order viscoelastic model. Considering the multiaxial condition, the stress–strain relation can be represented as follows:

$$\{\sigma\} = E(t)[C]\{\varepsilon_0\} + \int_0^t E(t-\xi)[C]\left\{\frac{d\varepsilon(\xi)}{d\xi}\right\}d\xi \tag{8.2}$$

in which $\{\sigma\}$ and $\{\varepsilon\}$ are stress and strain tensors in the vector form, respectively, and $[C]$ is the elastic matrix which is depended on Poisson's ratio. The elastic matrix for a two-dimensional problem can be determined for plane stress and plane strain conditions by using the following relations, respectively:

$$[C] = \frac{1}{1-\nu^2}\begin{bmatrix} 1 & \nu & 0 \\ \nu & 1 & 0 \\ 0 & 0 & (1-\nu)/2 \end{bmatrix} \tag{8.3}$$

$$[C] = \frac{1-\nu}{(1+\nu)(1-2\nu)}\begin{bmatrix} 1 & \dfrac{\nu}{1-\nu} & 0 \\ \dfrac{\nu}{1-\nu} & 1 & 0 \\ 0 & 0 & (1-2\nu)/[2(1-\nu)] \end{bmatrix} \tag{8.4}$$

It is evident that the abovementioned equations consider the Poisson's ratio as a constant parameter which may not always be true since the Poisson's ratio of a viscoelastic material generally is a time-dependent parameter. However, the Poisson's ratio was assumed constant during loading and unloading, that is, its variation versus time can be neglected. A detailed numerical solution was derived in Chapter 6, Interconversion of Constitutive Viscoelastic Functions, for the determination of time-dependent Poisson's ratio based on complex shear modulus as well as relaxation modulus of bitumen. Therefore if it is required to consider the variation of Poisson's ratio versus time, it will be needed to update elastic matrices in each time step of an incremental solution.

In general, the incremental form of viscoelastic relations is used in numerical methods where domain variables are known at $t = t_L$ and it is needed to determine within a time step $\Delta t_{L+1} = t_{L+1} - t_L$ under loading or unloading conditions by considering the variation of material properties. To derive an incremental formulation for a viscoelastic domain, it is required to find the relation of $\{\Delta\sigma_{L+1}\}$ and $\{\Delta\varepsilon_{L+1}\}$, which are incremental stress and incremental strain, respectively. For this purpose, by using Eq. (8.2), the stress tensor can be written for two successive time steps at t_L and t_{L+1} as follows:

$$\{\sigma_L\} = E(t_L)[C]\{\varepsilon_0\} + \int_0^{t_L} E(t_L - \xi)[C]\left\{\frac{d\varepsilon(\xi)}{d\xi}\right\}d\xi \tag{8.5}$$

$$\{\sigma_{L+1}\} = E(t_{L+1})[C]\{\varepsilon_0\} + \int_0^{t_{L+1}} E(t_{L+1} - \xi)[C]\left\{\frac{d\varepsilon(\xi)}{d\xi}\right\}d\xi \tag{8.6}$$

Knowing that $t_{L+1} = \Delta t_{L+1} + t_L$, it is possible to rewrite Eq. (8.6) as:

$$\{\sigma_{L+1}\} = E(t_{L+1})[C]\{\varepsilon_0\} + \int_0^{t_L+\Delta t_{L+1}} E((t_L + \Delta t_{L+1}) - \xi)[C]\left\{\frac{d\varepsilon(\xi)}{d\xi}\right\}d\xi \tag{8.7}$$

By subtracting Eq. (8.5) from Eq. (8.6), it is possible to determine $\{\Delta\sigma_{L+1}\} = \{\sigma_{L+1}\} - \{\sigma_L\}$ as:

$$\begin{aligned}\{\Delta\sigma_{L+1}\} = {} & \underbrace{E(t_{L+1})[C]\{\varepsilon_0\} - E(t_L)[C]\{\varepsilon_0\}}_{\text{I}} \\ & \underbrace{+ \int_0^{t_L+\Delta t_{L+1}} E((t_L + \Delta t_{L+1}) - \xi)[C]\left\{\frac{d\varepsilon(\xi)}{d\xi}\right\}d\xi}_{\text{II}} \\ & \underbrace{- \int_0^{t_L} E(t_L - \xi)[C]\left\{\frac{d\varepsilon(\xi)}{d\xi}\right\}d\xi}_{\text{III}}\end{aligned} \tag{8.8}$$

To facilitate the solving of Eq. (8.8), it is separated into three subequations labeled as (I), (II), and (III). By dividing the integration domain into two parts, it is possible to simplify the expression II as follows:

$$\begin{aligned}& \int_0^{t_L+\Delta t_{L+1}} E((t_L + \Delta t_{L+1}) - \xi)[C]\left\{\frac{d\varepsilon(\xi)}{d\xi}\right\}d\xi \\ & = \int_0^{t_L} E((t_L + \Delta t_{L+1}) - \xi)[C]\left\{\frac{d\varepsilon(\xi)}{d\xi}\right\}d\xi \\ & \quad + \int_{t_L}^{t_L+\Delta t_{L+1}} E((t_L + \Delta t_{L+1}) - \xi)[C]\left\{\frac{d\varepsilon(\xi)}{d\xi}\right\}d\xi\end{aligned} \tag{8.9}$$

In this stage, the Prony series, represented in Eq. (3.25) for the relaxation modulus, can be considered as a well-known mathematical representation of integer-order viscoelastic models. Furthermore, the amount of $d\varepsilon/dt$ is assumed as $\{\Delta\varepsilon_{L+1}/\Delta t_{L+1}\}$ for each time step of Δt_{L+1}. It is an important assumption to proceed with the derivation of an incremental form of viscoelastic relations, that should be checked carefully by considering the mechanical behavior of the desired material under boundary and loading conditions of the problem and loading history. For example, such an assumption is not true for high shear rate unless using very small time steps to ensure that there is no significant variation at both sides of the loading history in each time step. Fig. 8.1 shows this assumption for an arbitrary history of strain. Consequently, the second term in the right-hand side of Eq. (8.9) can be simply integrated as:

$$\int_{t_L}^{t_L+\Delta t_{L+1}} E((t_L+\Delta t_{L+1})-\xi)[C]\left\{\frac{d\varepsilon(\xi)}{d\xi}\right\}d\xi$$

$$=[C]\frac{\{\Delta\varepsilon_{L+1}\}}{\Delta t_{L+1}}\int_{t_L}^{t_L+\Delta t_{L+1}} E((t_L+\Delta t_{L+1})-\xi)\,d\xi \qquad (8.10)$$

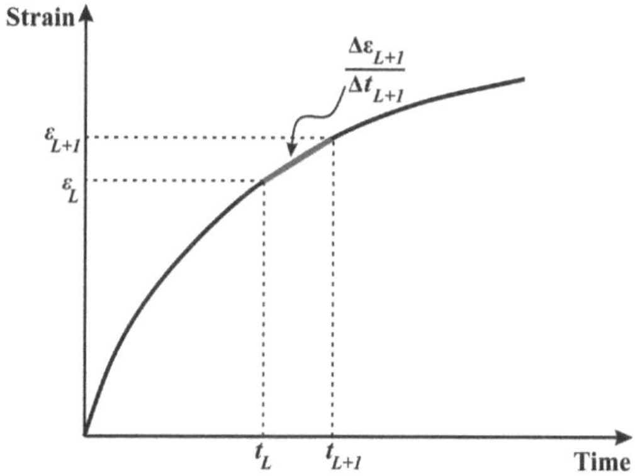

Figure 8.1 The main assumption for strain variation in each time step to derive incremental form of viscoelastic relation.

and then, the integral in Eq. (8.10) is calculated by using the properties of the exponential function as follows:

$$
\begin{aligned}
&\int_{t_L}^{t_L+\Delta t_{L+1}} E((t_L+\Delta t_{L+1})-\xi)\,d\xi \\
&= \int_{t_L}^{t_L+\Delta t_{L+1}} \left(E_\infty + \sum_{m=1}^{M} E_m \exp\left((\xi-(t_L+\Delta t_{L+1}))/\tau_m\right)\right) d\xi \\
&= E_\infty \xi + \sum_{m=1}^{M} E_m \tau_m \exp\left((\xi-(t_L+\Delta t_{L+1}))/\tau_m\right)\Big|_{t_L}^{t_L+\Delta t_{L+1}} \\
&= E_\infty \Delta t_{L+1} + \sum_{m=1}^{M} E_m \tau_m \left[1-\exp\left(-\Delta t_{L+1}/\tau_m\right)\right]
\end{aligned} \tag{8.11}
$$

Substituting Eq. (8.11) into Eq. (8.10) resulted in the following equation:

$$
\int_{t_L}^{t_L+\Delta t_{L+1}} E((t_L+\Delta t_{L+1})-\xi)[C]\left\{\frac{d\varepsilon(\xi)}{d\xi}\right\}d\xi = \tilde{E}[C]\{\Delta\varepsilon_{L+1}\} \tag{8.12}
$$

in which

$$
\tilde{E} = E_\infty + \sum_{m=1}^{M} \frac{E_m \tau_m}{\Delta t_{L+1}}\left[1-\exp\left(-\Delta t_{L+1}/\tau_m\right)\right] \tag{8.13}
$$

The remaining term of the expression II in Eq. (8.8), which is not determined up to now, has the same integration range with expression III and can be directly added to this expression. The result of this summation would be:

$$
\int_0^{t_L} \left[E((t_L+\Delta t_{L+1})-\xi) - E(t_L-\xi)\right][C]\left\{\frac{d\varepsilon(\xi)}{d\xi}\right\}d\xi \tag{8.14}
$$

To calculate Eq. (8.14), it is needed to calculate the following expression first:

$$
E((t_L+\Delta t_{L+1})-\xi) - E(t_L-\xi) = \sum_{m=1}^{M} E_m \tau_m \exp(\xi - t_L)\left[\exp\left(-\Delta t_{L+1}/\tau_m\right)-1\right] \tag{8.15}
$$

and then by substituting Eq. (8.15) into Eq. (8.14), it would be:

$$
\begin{aligned}
&\int_0^{t_L} \left[E((t_L+\Delta t_{L+1})-\xi) - E(t_L-\xi)\right][C]\left\{\frac{d\varepsilon(\xi)}{d\xi}\right\}d\xi \\
&= \sum_{m=1}^{M} \left[\exp\left(-\Delta t_{L+1}/\tau_m\right)-1\right]\left\{\alpha_m^L\right\}
\end{aligned} \tag{8.16}
$$

in which,

$$\{\alpha_m^L\} = \int_0^{t_L} E_m \exp\left(\frac{\xi - t_L}{\tau_m}\right)[C]\left\{\frac{d\varepsilon(\xi)}{d\xi}\right\}d\xi \tag{8.17}$$

By implementing a procedure similar to the one used for determination of a part of expression II, the integration range should be divided into two parts and, considering the relation of time step $\Delta t_L = t_L - t_{L-1}$, we have:

$$\{\alpha_m^L\} = \int_0^{t_{L-1}} E_m \exp\left(\frac{\xi - t_L}{\tau_m}\right)[C]\left\{\frac{d\varepsilon(\xi)}{d\xi}\right\}d\xi + \int_{t_{L-1}}^{t_L} E_m \exp\left(\frac{\xi - t_L}{\tau_m}\right)[C]\left\{\frac{d\varepsilon(\xi)}{d\xi}\right\}d\xi \tag{8.18}$$

The second term on the right-hand side of Eq. (8.18) can be approximately determined by assuming a constant value of $\Delta\varepsilon/\Delta t_L$ for $d\varepsilon/dt$ within the time step of Δt_L:

$$\int_{t_{L-1}}^{t_L} E_m \exp\left(\frac{\xi - t_L}{\tau_m}\right)[C]\left\{\frac{d\varepsilon(\xi)}{d\xi}\right\}d\xi = \frac{E_m \tau_m}{\Delta t_L}\left[1 - \exp\left(-\Delta t_L/\tau_m\right)\right][C]\{\Delta\varepsilon_L\} \tag{8.19}$$

With regard to Eq. (8.17):

$$\{\alpha_m^{L-1}\} = \int_0^{t_{L-1}} E_m \exp\left(\frac{\xi - t_{L-1}}{\tau_m}\right)[C]\left\{\frac{d\varepsilon(\xi)}{d\xi}\right\}d\xi \tag{8.20}$$

Then, the first term on the right-hand side of Eq. (8.18) is determined by considering the relation of time step $\Delta t_L = t_L - t_{L-1}$ as:

$$\int_0^{t_{L-1}} E_m \exp\left(\frac{\xi - t_L}{\tau_m}\right)[C]\left\{\frac{d\varepsilon(\xi)}{d\xi}\right\}d\xi = \exp\left(-\Delta t_L/\tau_m\right)\{\alpha_m^{L-1}\} \tag{8.21}$$

Finally, with regard to Eqs. (8.18) to (8.21), $\{\alpha_m^L\}$ is calculated as:

$$\{\alpha_m^L\} = \exp\left(-\Delta t_L/\tau_m\right)\{\alpha_m^{L-1}\} + \frac{E_m \tau_m\left[1 - \exp\left(-\Delta t_L/\tau_m\right)\right]}{\Delta t_L}[C]\{\Delta\varepsilon_L\} \tag{8.22}$$

Therefore it can be seen that Eq. (8.15) is determined now and with the expression I [in Eq. (8.8)] is called "residual stress," $\{\sigma_{\text{res}}\}$, and is presented as follows:

$$\{\sigma_{\text{res}}\} = [E(t_{L+1}) - E(t_L)]\,[C]\,\{\varepsilon(0)\} + \sum_{m=1}^{M}\left[\exp\left(-\Delta t_{L+1}/\tau_m\right) - 1\right]\,\{\alpha_m^L\} \tag{8.23}$$

It is necessary to mention that $\{\alpha_m^0\} = 0$ based on Eq. (8.17). Having all expressions (I, II, and III) in Eq. (8.8), $\{\Delta\sigma_{L+1}\}$ is consequently determined as follows:

$$\{\Delta\sigma_{L+1}\} = \tilde{E}[C]\{\Delta\varepsilon_{L+1}\} + \{\sigma_{\text{res}}\} \tag{8.24}$$

The residual stress in Eq. (8.24) shows the significant difference between an elastic and a viscoelastic solution. It can be seen that applying strain at each time step resulted in the generation of residual stress which will affect the next time step.

8.2.2 Fractional-order viscoelastic models

8.2.2.1 Binomial expansion approach

As mentioned in Section 3.3, fractional-order viscoelastic models are another category to describe the mechanical behavior of viscoelastic materials which have some advantages to encourage their use instead of integer-order viscoelastic models for some cases. In this section, the incremental viscoelastic relations are developed by considering the simple fractional element (see Section 3.3.2) (Hajikarimi, Onochie, & Fini, 2020). Having a power relation for relaxation modulus as mentioned in Eq. (3.51) with a general form of $At^{-\beta}$ in which $A = E\tau^{\beta}/\Gamma(1-\beta)$ (Hajikarimi, Aflaki, & Hoseini, 2013) and also assuming that $d\varepsilon/dt = \Delta\varepsilon/\Delta t$ in the time step of $\Delta t_{R+1} = t_{R+1} - t_R$, the second expression of the right-hand side of Eq. (8.9) can be written as follows:

$$\int_{t_R}^{t_R+\Delta t_{R+1}} E((t_R + \Delta t_{R+1}) - \xi)[C]\left\{\frac{d\varepsilon(\xi)}{d\xi}\right\}d\xi$$

$$= [C]\frac{\{\Delta\varepsilon_{R+1}\}}{\Delta t_{R+1}}\int_{t_R}^{t_R+\Delta t_{R+1}} E((t_R + \Delta t_{R+1}) - \xi)\,d\xi \tag{8.25}$$

The integral of Eq. (8.25) can be simply determined considering the abovementioned assumptions:

$$\int_{t_R}^{t_R+\Delta t_{R+1}} E((t_R + \Delta t_{R+1}) - \xi)d\xi = \int_{t_R}^{t_R+\Delta t_{R+1}} A((t_R+\Delta t_{R+1})-\xi)^{-\beta}d\xi$$

$$= A\frac{((t_R+\Delta t_{R+1})-\xi)^{1-\beta}}{\beta-1}\Bigg|_{t_R}^{t_R+\Delta t_{R+1}} = A\frac{(\Delta t_{R+1})^{1-\beta}}{1-\beta} \tag{8.26}$$

and substituting Eq. (8.26) into Eq. (8.25):

$$\int_{t_R}^{t_R+\Delta t_{R+1}} E((t_R + \Delta t_{R+1}) - \xi)[C]\left\{\frac{d\varepsilon(\xi)}{d\xi}\right\}d\xi = \tilde{E}[C]\{\Delta\varepsilon_{R+1}\} \tag{8.27}$$

in which:

$$\tilde{E} = A\frac{(\Delta t_{R+1})^{-\beta}}{1-\beta} \tag{8.28}$$

The remaining part of the expression II of Eq. (8.9) has an equivalent domain with the expression III of Eq. (8.8) and can be directly added to it:

$$\int_0^{t_R} [E((t_R + \Delta t_{R+1}) - \xi) - E(t_R - \xi)]\,[C]\left\{\frac{d\varepsilon(\xi)}{d\xi}\right\}d\xi \tag{8.29}$$

To determine Eq. (8.29), first it is necessary to calculate the following equation:

$$E((t_R + \Delta t_{R+1}) - \xi) - E(t_R - \xi) = A\left[((t_R+\Delta t_{R+1})-\xi)^{-\beta} - (t_R-\xi)^{-\beta}\right] \tag{8.30}$$

A binomial in the form of $(A + B)^p$ can be expanded by using the following relation:

$$(A+B)^p = \sum_{k=0}^{\infty} C(p,k)A^{p-k}B^k \tag{8.31}$$

in which for $k>0$

$$C(p,k) = \frac{p(p-1)(p-2)\ldots(p-k+1)}{k!} \tag{8.32}$$

Using Eq. (8.31), Eq. (8.30) can be rewritten as:

$$\begin{aligned} & A\left[((t_R+\Delta t_{R+1})-\xi)^{-\beta}-(t_R-\xi)^{-\beta}\right] \\ & = A\,(t_R-\xi)^{-\beta}+A\sum_{k=1}^{\infty}C(-\beta,\ k)(t_R-\xi)^{-\beta-k}(\Delta t_{R+1})^k - A\,(t_R-\xi)^{-\beta} \\ & = A\sum_{k=1}^{\infty}C(-\beta,\ k)(t_R-\xi)^{-\beta-k}(\Delta t_{R+1})^k \end{aligned} \tag{8.33}$$

Substituting Eq. (8.33) into Eq. (3.29):

$$\begin{aligned} & \int_0^{t_R}[E((t_R+\Delta t_{R+1})-\xi)-E(t_R-\xi)]\,[C]\left\{\frac{d\varepsilon(\xi)}{d\xi}\right\}d\xi \\ & = A\sum_{k=1}^{\infty}C(-\beta,\ k)\,(\Delta t_{R+1})^k\int_0^{t_R}(t_R-\xi)^{-\beta-k}\,[C]\left\{\frac{d\varepsilon(\xi)}{d\xi}\right\}d\xi \end{aligned} \tag{8.34}$$

in which:

$$\{\alpha_k^R\}=\int_0^{t_R}(t_R-\xi)^{-\beta-k}\,[C]\left\{\frac{d\varepsilon(\xi)}{d\xi}\right\}d\xi \tag{8.35}$$

By splitting the integration domain of Eq. (8.35) to achieve a recursive formula for α_k^R:

$$\begin{aligned} \{\alpha_k^R\} = & \int_0^{t_{R-1}}(t_R-\xi)^{-\beta-k}\,[C]\left\{\frac{d\varepsilon(\xi)}{d\xi}\right\}d\xi \\ & +\int_{t_{R-1}}^{t_R}(t_R-\xi)^{-\beta-k}\,[C]\left\{\frac{d\varepsilon(\xi)}{d\xi}\right\}d\xi \end{aligned} \tag{8.36}$$

The second expression on the right-hand side of Eq. (8.36) can be simply determined by assuming $d\varepsilon/dt=\Delta\varepsilon/\Delta t$ in the time step of Δt_R as follows:

$$\begin{aligned} & \int_{t_{R-1}}^{t_R}(t_R-\xi)^{-\beta-k}\,[C]\left\{\frac{d\varepsilon(\xi)}{d\xi}\right\}d\xi \\ & = \int_{t_{R-1}}^{t_R}((t_{R-1}+\Delta t_R)-\xi)^{-\beta-k}\,[C]\left\{\frac{d\varepsilon(\xi)}{d\xi}\right\}d\xi \\ & = \frac{(\Delta t_R)^{-\beta-k}}{1-\beta-k}\,[C]\,\{\Delta\varepsilon_R\} \end{aligned} \tag{8.37}$$

And the first expression on the right-hand side of Eq. (8.36) with regard to the following relation:

$$\{\alpha_k^{R-1}\}=\int_0^{t_{R-1}}(t_{R-1}-\xi)^{-\beta-k}\,[C]\left\{\frac{d\varepsilon(\xi)}{d\xi}\right\}d\xi \tag{8.38}$$

and knowing that $\Delta t_R = t_R - t_{R-1}$ will be determined as:

$$\int_0^{t_{R-1}} ((t_{R-1}-\xi)+\Delta t_R)^{-\beta-k}[C]\left\{\frac{d\varepsilon(\xi)}{d\xi}\right\}d\xi$$

$$=\sum_{r=0}^{\infty} C(-\beta-k,\ r)\,(\Delta t_R)^r \int_0^{t_{R-1}} (t_{R-1}-\xi)^{-\beta-k-r}[C]\left\{\frac{d\varepsilon(\xi)}{d\xi}\right\}d\xi \quad (8.39)$$

Thus α_k^R can be derived as:

$$\{\alpha_k^R\} = \{\alpha_k^{R-1}\} + \sum_{r=1}^{\infty} C(-\beta-k,\ r)\,(\Delta t_R)^r \int_0^{t_{R-1}} (t_{R-1}-\xi)^{-\beta-k-r}[C]\left\{\frac{d\varepsilon(\xi)}{d\xi}\right\}d\xi$$

$$+\frac{(\Delta t_R)^{-\beta-k}}{1-\beta-k}[C]\,\{\Delta\varepsilon_R\} \quad (8.40)$$

with $\{\alpha_k{}^0\} = 0$.

According to Eq. (8.8), it is evident that:

$$\{\Delta\sigma_{R+1}\} = \tilde{E}\,[C]\,\{\Delta\varepsilon_{R+1}\} + \{\sigma_{res}\} \quad (8.41)$$

in which $\tilde{E}$ is equivalent modulus and σ_{res} is residual stress for the current time step which can be determined as:

$$\{\sigma_{res}\} = [E(t_{R+1}) - E(t_R)]\,[C]\,\{\varepsilon(0)\} + A\sum_{k=1}^{\infty} C(-\beta, k)\,(\Delta t_{R+1})^k\,\{\alpha_k^R\} \quad (8.42)$$

8.2.2.2 Taylor expansion and Gauss–Legendre method

Recalling Eq. (8.8), it is possible to split the integral domain of the second term on the right-hand side of this equation and add the integral of the third term (with an integral range from 0 to t_L) to achieve the following equation as:

$$\{\Delta\sigma_{L+1}\} = \underbrace{[E(t_{L+1}) - E(t_L)][C]\{\varepsilon_0\}}_{\text{I}}$$

$$+\underbrace{\int_0^{t_L} [E((t_L+\Delta t_{L+1})-\xi)][C]\left\{\frac{d\varepsilon(\xi)}{d\xi}\right\}d\xi}_{\text{II}}$$

$$+\underbrace{\int_0^{t_L+\Delta t_{L+1}} E(t_L+\Delta t_{L+1}-\xi)[C]\left\{\frac{d\varepsilon(\xi)}{d\xi}\right\}d\xi}_{\text{III}} \quad (8.43)$$

The first and third terms of Eq. (8.43) can be simply determined considering $E(t) = At^{-\beta}$ as:

$$\underbrace{[E(t_{L+1}) - E(t_L)][C]\{\varepsilon_0\}}_{\text{I}} = A \times [C] \times \{\varepsilon_0\} \times \left[(t_{L+1})^{-\beta} - (t_L)^{-\beta}\right] \quad (8.44)$$

and,

$$\int_{t_L}^{t_L+\Delta t_{L+1}} E(t_L + \Delta t_{L+1} - \xi)[C]\left\{\frac{d\varepsilon(\xi)}{d\xi}\right\}d\xi$$

$$= \frac{1}{1-\beta} \times A \times [C] \times \{\Delta\varepsilon_{L+1}\} \times (\Delta t_{L+1})^{-\beta} \quad (8.45)$$

To determine the second term on the right-hand side of Eq. (8.43), the Taylor expansion is used here. Assuming small time variation, it is possible to use the linear approximation of the Taylor expansion as:

$$((t_L + \Delta t_{L+1}) - \xi)^{-\beta} = (t_L - \xi)^{-\beta} + \beta \times \Delta t_{L+1} \times (t_L - \xi)^{-\beta-1} \quad (8.46)$$

in the following equation:

$$\int_0^{t_L} [E((t_L + \Delta t_{L+1}) - \xi) - E(t_L - \xi)][C]\left\{\frac{d\varepsilon(\xi)}{d\xi}\right\}d\xi$$

$$= \int_0^{t_L} A\left[((t_L + \Delta t_{L+1}) - \xi)^{-\beta} - (t_L - \xi)^{-\beta}\right][C]\left\{\frac{d\varepsilon(\xi)}{d\xi}\right\}d\xi \quad (8.47)$$

would be,

$$\int_0^{t_L} A\left[(t_L - \xi)^{-\beta} + \beta \times \Delta t_{L+1} \times (t_L - \xi)^{-\beta-1} - (t_L - \xi)^{-\beta}\right][C]\left\{\frac{d\varepsilon(\xi)}{d\xi}\right\}d\xi$$

$$= A \times \beta \times \Delta t_{L+1}[C]\int_0^{t_L} \left[(t_L - \xi)^{-\beta-1}\right]\left\{\frac{d\varepsilon(\xi)}{d\xi}\right\}d\xi \quad (8.48)$$

The integral of Eq. (8.48) cannot be calculated individually due to the presence of $d\varepsilon/dt$. Therefore it is necessary to simplify this integration by using numerical methods. For this purpose, the remained integral is divided into two parts and it is tried to eliminate $d\varepsilon/dt$ within this integral by using the concept of integration by parts. Eqs. (8.49) to (8.53) shows these calculations:

$$\alpha_m^L = \int_0^{t_L} \left[(t_L-\xi)^{-\beta-1}\right] \left\{\frac{d\varepsilon(\xi)}{d\xi}\right\} d\xi = \int_0^{t_{L-1}} \left[(t_L-\xi)^{-\beta-1}\right] \left\{\frac{d\varepsilon(\xi)}{d\xi}\right\} d\xi + \int_{t_{L-1}}^{t_L} \left[(t_L-\xi)^{-\beta-1}\right] \left\{\frac{d\varepsilon(\xi)}{d\xi}\right\} d\xi \tag{8.49}$$

The first term on the right-hand side of Eq. (8.49) can be written as:

$$\int_0^{t_{L-1}} \left[(t_L-\xi)^{-\beta-1}\right] \left\{\frac{d\varepsilon(\xi)}{d\xi}\right\} d\xi = \int u \times dv = u \times v - \int v \times du$$

$$u = \left[(t_L-\xi)^{-\beta-1}\right] \quad \rightarrow \quad du = (\beta+1)\left[(t_L-\xi)^{-\beta-2}\right] d\xi$$

$$dv = \left\{\frac{d\varepsilon(\xi)}{d\xi}\right\} d\xi \quad \rightarrow \quad v = \int \left\{\frac{d\varepsilon(\xi)}{d\xi}\right\} d\xi = \varepsilon(\xi) \tag{8.50}$$

and determined as follows:

$$\left[(t_L-\xi)^{-\beta-1}\varepsilon(\xi)\right]_0^{t_{L-1}} - (\beta+1)\int_0^{t_{L-1}} \left[(t_L-\xi)^{-\beta-2}\right] \times \varepsilon(\xi) d\xi$$

$$= (\Delta t_L)^{-\beta-1}\varepsilon_{L-1} - (\beta+1)\int_0^{t_{L-1}} \left[(t_L-\xi)^{-\beta-2}\right] \times \varepsilon(\xi) d\xi \tag{8.51}$$

The second term on the right-hand side of Eq. (8.49) is also determined as:

$$\int_{t_{L-1}}^{t_L} \left[(t_L-\xi)^{-\beta-1}\right] \left\{\frac{d\varepsilon(\xi)}{d\xi}\right\} d\xi = -\frac{1}{\beta} \times \frac{\Delta\varepsilon_L}{\Delta t_L} \times (\Delta t_L)^{-\beta}$$

$$= -\frac{1}{\beta} \times \Delta\varepsilon_L \times \Delta t_L{}^{-\beta-1} \tag{8.52}$$

and finally, $\{\alpha_m^L\}$ would be,

$$\alpha_m^L = (\Delta t_L)^{-\beta-1}\varepsilon_{L-1} - (\beta+1)\int_0^{t_{L-1}} \left[(t_L-\xi)^{-\beta-2}\right] \times \varepsilon(\xi) d\xi - \frac{1}{\beta} \times \Delta\varepsilon_L \times \Delta t_L{}^{-\beta-1} \tag{8.53}$$

The integral derived in Eq. (8.53) has no $d\varepsilon/dt$ and can be approximated by implementing the Gauss–Legendre method. For this purpose, at first it is needed to map the integration range to $[-1, +1]$ and then use the numerical Gauss–Legendre method. The mapping procedure is presented in Eq. (8.54) as:

$$\left.\begin{array}{l}\Psi = \int_0^{t_{L-1}} \left[(t_L-\xi)^{-\beta-2}\right] \times \varepsilon(\xi)d\xi \\ \xi = \dfrac{t_{L-1} \times x + t_{L-1}}{2} \rightarrow d\xi = \dfrac{t_{L-1}}{2}dx \end{array}\right\} \rightarrow \Psi$$

$$= \int_{-1}^{1} \left[\left(t_L - \frac{t_{L-1} \times x + t_{L-1}}{2}\right)^{-\beta-2}\right] \times \varepsilon\left(\frac{t_{L-1} \times x + t_{L-1}}{2}\right) \times \frac{t_{L-1}}{2} dx \tag{8.54}$$

which can be calculated by using the following equation:

$$\Psi \approx \sum_{i=1}^{15} w_i F(x_i) \tag{8.55}$$

in which:

$$F(x_i) = \frac{t_{L-1}}{2} \times \left[\left(t_L - \frac{t_{L-1} \times x_i + t_{L-1}}{2}\right)^{-\beta-2} \times \varepsilon\left(\frac{t_{L-1} \times x_i + t_{L-1}}{2}\right)\right] \tag{8.56}$$

where x_i and w_i are roots and weights of the Legendre equation which are presented in Table 8.1 for $n = 15$.

Table 8.1 Roots and weights of the Legendre equation for $n = 15$.

i	Weight, w_i	Abscissa, x_i
1	0.2025782419255613	0.0000000000000000
2	0.1984314853271116	−0.2011940939974345
3	0.1984314853271116	0.2011940939974345
4	0.1861610000155622	−0.3941513470775634
5	0.1861610000155622	0.3941513470775634
6	0.1662692058169939	−0.5709721726085388
7	0.1662692058169939	0.5709721726085388
8	0.1395706779261543	−0.7244177313601701
9	0.1395706779261543	0.7244177313601701
10	0.1071592204671719	−0.8482065834104272
11	0.1071592204671719	0.8482065834104272
12	0.0703660474881081	−0.9372733924007060
13	0.0703660474881081	0.9372733924007060
14	0.0307532419961173	−0.9879925180204854
15	0.0307532419961173	0.9879925180204854

Consequently, the stress variation is determined as follows:

$$\{\Delta\sigma_{L+1}\} = A \times [C] \times \{\varepsilon_0\} \times \left((t_{L+1})^{-\beta} - (t_L)^{-\beta}\right) + \frac{1}{1-\beta} \times A \times [C]$$
$$\times \{\Delta\varepsilon_{L+1}\} \times (\Delta t_{L+1})^{-\beta} + + A \times \beta \times \Delta t_{L+1}[C]$$
$$\times \left((\Delta t_L)^{-\beta-1}\left(\varepsilon_{L-1} - \frac{\Delta\varepsilon_L}{\beta}\right) - (\beta+1)\sum_{i=1}^{15} w_i F_{(x_i)}\right) \quad (8.57)$$

in which

$$F(x_i) = \frac{t_{L-1}}{2} \times \left[\left(t_L - \frac{t_{L-1} \times x_i + t_{L-1}}{2}\right)^{-\beta-2} \times \varepsilon\left(\frac{t_{L-1} \times x_i + t_{L-1}}{2}\right)\right] \quad (8.58)$$

where x_i is the root of Legendre polynomial and w_i is the Gauss–Legendre weight factor, as found in Table 8.1 for $n = 15$.

8.3 Incremental form of viscoelastic relations in the finite element framework

Using the abovementioned incremental form of integer-order or fractional-order viscoelastic models, FEM codes can be developed in MATLAB, FORTRAN, and other programming languages to construct and analyze the desired models of viscoelastic materials. To derive the discretized force–displacement equation, it is well-known that the FE approximation in two dimensions is as follows:

$$\{u\} = [N]\{U\} \quad (8.59)$$

in which $\{U\}$ is all the nodal degree of freedoms associated with the element and $[N]$ is the matrix of shape function which can be expressed as:

$$[N] = [N_1 \quad N_2 \quad \ldots \quad N_I \quad \ldots] \quad (8.60)$$

where N_I belongs to the node I as:

$$\mathrm{N}_I = \begin{bmatrix} N_I & 0 \\ 0 & N_I \end{bmatrix} \quad (8.61)$$

in which N_I is directly depended on the type of element used for a specific problem. Generally, for an element e, the incremental form of virtual work principle is represented as follows:

$$\int_{\Omega^e} \delta\{\Delta\varepsilon\}^T\{\Delta\sigma\}d\Omega = \int_{\Omega^e} \delta\{\Delta u\}^T\{\Delta b\}d\Omega + \int_{\Gamma^e} \delta\{\Delta u\}^T\{\Delta p\}d\Gamma \quad (8.62)$$

in which δ is the first-order derivation operator; $\{\ \}^T$ is the matrix transpose operator; and $\{\Delta u\}$, $\{\Delta p\}$, and $\{\Delta b\}$ are the incremental displacement, the prescribed traction, and the body force, respectively. According to Eq. (8.59), it is evident that the incremental form of this approximation would be:

$$\{\Delta u\} = [N]\{\Delta U\} \quad (8.63)$$

On the other hand, by assuming small deformation and using the strain—displacement relation, the following equation can be written:

$$\{\Delta\varepsilon\} = [B]\{\Delta U\} \quad (8.64)$$

in which $[B]$ is the strain—displacement matrix as:

$$[B] = \begin{bmatrix} B_1 & B_2 & \ldots & B_I & \ldots \end{bmatrix} \quad (8.65)$$

where B_I belongs to node I and can be expressed as follows:

$$B_I = \begin{bmatrix} N_{I,x} & 0 \\ 0 & N_{I,y} \\ N_{I,y} & N_{I,x} \end{bmatrix} \quad (8.66)$$

The residual force, F_{res}, should be determined from the residual stress, σ_{res}. Based on the abovementioned general equations for a FE framework, Zhang and Li (2009) derived the following equations for analyzing a linear viscoelastic domain numerically:

$$[K]\{\Delta U\} = \{\Delta F\} + \{F_{\text{res}}\} \quad (8.67)$$

in which

$$[K] = \int_{\Omega} \tilde{E}[B]^T[C][B]d\Omega \quad (8.68)$$

$$\{\Delta F\} = \int_{\Omega^e} [N]^T\{\Delta b_{R+1}\}d\Omega + \int_{\Psi^e} [N]^T\{\Delta p_{R+1}\}d\Psi \quad (8.69)$$

$$\{F_{\text{res}}\} = -\int_{\Omega^e} [B]^T\{\sigma_{\text{res}}\}d\Omega \quad (8.70)$$

where $[K]$ is the equivalent stiffness matrix, $\{\Delta F\}$ is the equivalent incremental load in the interval Δt_{L+1}, and $\{F_{\text{res}}\}$ is the residual force due to the residual stress. Fig. 8.2 depicts a flowchart proposed by Zhang and Li (2009) to solve viscoelastic problems by using an FE framework as well as an XFEM which will be described in the following section.

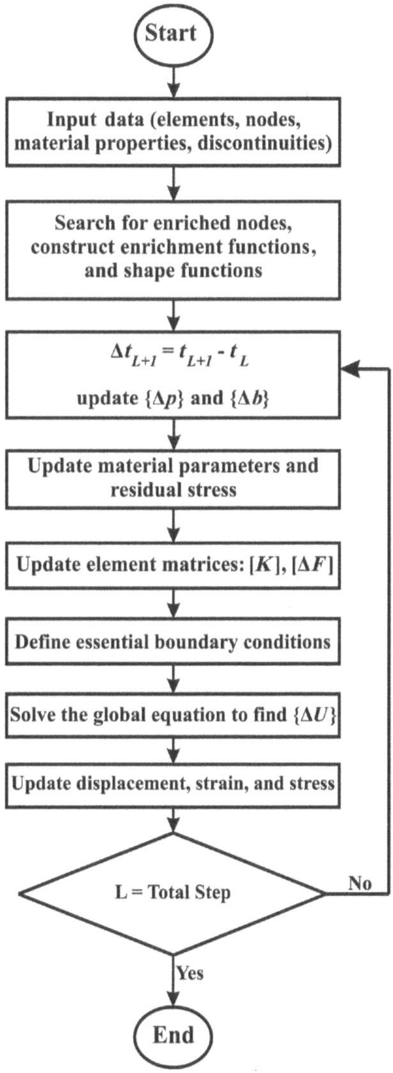

Figure 8.2 Flowchart for solving viscoelastic problems by using finite element method.

8.4 Extended finite element method for a cracked linear viscoelastic medium

To better understand how the proposed flowchart shown in Fig. 8.2 can be applied in an FE framework, a section of a code to calculate stiffness matrix developed for an XFEM analysis of a two-dimensional cracked linear viscoelastic medium in MATLAB is presented in Box 8.1 (Hajikarimi et al., 2012) as a general code which can be implemented for both continuous and discontinuous media. The main advantage of XFEM compared to FEM is that in this method the mesh is not dependent of the internal boundaries such as inclusion boundaries or cracks interfaces. Therefore no mesh updating is required to solve the problems by implementing the XFEM (Moës et al., 1999). Considering such an advantage, the XFEM is widely used to simulate crack propagation problems and fracture analysis of composites (Mohammadi, 2012). Fig. 8.3 illustrates a comparison between XFEM and FEM considering an inclusion within a viscoelastic domain. As can be seen in this figure, by using the XFEM, a structured mesh can be used to apply FE relations.

The following equation represents the approximation in element e of a cracked viscoelastic domain:

$$u(x) = \sum_{I \in C_A} N_I(x)u_I + \sum_{J \in C_S} N_J(x)H_J(x)a_J + \sum_{K \in C_T} N_K(x)\left(\sum_{m=1}^{4} \Phi_K^m(x)c_K^m\right) \tag{8.71}$$

where C_A, C_S, and C_T are the sets of conventional nodes, split nodes, and tip nodes, respectively. The split nodes are associated with elements containing crack surfaces and the tip nodes are associated with the element containing the crack tip, as illustrated in Fig. 8.4. The nodes associated with split and tip elements are called enriched nodes as some enrichment functions are used to enrich the shape functions used for such elements to omit the necessity to mesh updating. $N_I(x)$ is the shape function associated with the conventional node I. u_I, a_J, and c_K^m are conventional degrees of freedom, enriched nodes of split elements, and enriched nodes of the tip element, respectively. The $H(x)$ is a function used to reflect the discontinuity of displacement across a crack surface which can be usually taken as a Heaviside function as follows (Moës et al., 1999):

$$H(x) = \begin{cases} 1 & \text{if } x \text{ above the crack} \\ -1 & \text{if } x \text{ below the crack} \end{cases} \tag{8.72}$$

BOX 8.1 A section of a code developed for an extended finite element method analysis of a cracked linear viscoelastic medium in MATLAB.

```
% --------------------------------------------
% Initialize stiffness matrix, force vector

% Each split node is enriched by ONE function, H(x)
% Each tip node is enriched by FOUR functions, B_i(x)
% then, the total dofs is :
% total dof = numnode*nsd + numsplitnode*1*nsd + numstipnode*4*nsd
% here, two dimension, nsd = 2

total_unknown = numnode*2 + size(split_nodes,1)*1*2 + ...
    size(tip_nodes,1)*4*2;

K = sparse(total_unknown,total_unknown) ;
f = zeros(total_unknown,1) ;
f0 = zeros(total_unknown,1) ;
% --------------------------------------------

% **********************************
%    Stiffness matrix computation
% **********************************

% Compliance matrix C
if ( strcmp(stressState,'PLANE_STRESS') )
    C = 1/(1-nu^2)*[ 1   nu 0;
        nu  1  0 ;
        0   0  0.5*(1-nu) ];
else
    C = 1/(1+nu)/(1-2*nu)*[ 1-nu  nu  0;
        nu    1-nu 0;
        0     0  0.5-nu ];
end

% --------------------------------------------
% Loop on elements for determining Gauss Points
% --------------------------------------------

% ---------------------
%    NODAL FORCE VECTOR
% ---------------------

% ------------------------------------
% Loop on elements for each time step
% ------------------------------------

% sigma0 is a vector which contains residual stresses
% alpham is a vector which is needed for calculating residual stress

    alpham01 = zeros(3,numelem) ;
    alpham02 = zeros(3,numelem) ;
    sigma0L = zeros(3,numelem) ;
    u = zeros(total_unknown,1) ;
    utotal = zeros(total_unknown,1) ;
    stressPoints = [-1 -1;1 -1;1 1;-1 1];
    totalstress = zeros (numelem,4,3) ;
    totalstress2 = zeros (numelem,4,3) ;
    totalstress3 = zeros (numelem,4,3) ;
    uplot = zeros(Time/dt,1) ;
```

(Continued)

BOX 8.1 (Continued)

```
for L = 1 : (Time/dt)
    tL = L * dt;
    Ebar = Einf + (E1*to1/dt)*(1-exp(-dt/to1)) + (E2*to2/dt)*(1-exp(-dt/to2)) ;

f0 = zeros(total_unknown,1) ;
K = zeros(total_unknown,total_unknown) ;


for iel = 1 : numelem
sctr = element(iel,:); % element connectivity
nn   = length(sctr);   % number of nodes per element
U    = element_disp(iel,pos,enrich_node,u);
ke   = 0 ;             % elementary stiffness matrix
xTip=xTip2;
xCr=xCr2;
alpha=alpha2;
if (ismember(iel,split_elem))     % split element
    order = 2 ;
    phi   = ls2(sctr,1);
    [W,Q] = discontQ4quad(order,phi);
elseif (ismember(iel,tip_elem))   % tip element
    order = 7;
    phi = ls2(sctr,1);
    for inod=1:1
        if (EnodTpSpRef(sctr(inod))==1)
           xTip=xTip1;
           xCr=xCr1;
           alpha=alpha1;
        else
           xTip=xTip2;
           xCr=xCr2;
           alpha=alpha2;
        end
    end
    nodes = node(sctr,:);
    [W,Q] = disTipQ4quad(order,phi,nodes,xTip);
elseif ( any(intersect(tip_nodes,sctr)) ~= 0)% having tip enriched nodes
    order = 4 ;
    for inod=1:nn
        if (EnodTpSpRef(sctr(inod))==2)
           xTip=xTip2;
           xCr=xCr2;
           alpha=alpha2;
           break
        elseif (EnodTpSpRef(sctr(inod))==1)
           xTip=xTip1;
           xCr=xCr1;
           alpha=alpha1;
           break
        end
    end
    [W,Q] = quadrature(order,'GAUSS',2);

else
    order = 2 ;
    [W,Q] = quadrature(order,'GAUSS',2);
end
% ------------------------------------------------

% Determine the position in the global matrix K
sctrB = assembly(iel,enrich_node,pos);
% --------------------
% Loop on Gauss points
% --------------------
```

(Continued)

BOX 8.1 (Continued)

```
    for kk = 1 : size(W,1)
        pt = Q(kk,:);                                   % quadrature point
        % B matrix
        [B,J0] = private_xfemBmatrix4(pt,elemType,iel,enrich_node,...
                                      EnodTpSpRef,xCr,xTip,alpha);
        % Stiffness matrix
        % ke = ke + B'*C*B*W(kk)*det(J0);

        K(sctrB,sctrB) = K(sctrB,sctrB) +
thick*Ebar*B'*C*B*W(kk)*det(J0);

        strain = B * U ;

        alpham1L(:,iel) = exp(-dt/to1)*alpham01(:,iel) + (E1*to1*(1-
exp(-dt/to1))/dt)* C * strain ;

        alpham2L(:,iel) = exp(-dt/to2)*alpham02(:,iel) + (E2*to2*(1-
exp(-dt/to2))/dt)* C * strain ;

        sigma0L (:,iel) =  ( E1 * (exp(-tL/to1)-exp(-(tL-dt)/to1)) + E2
* (exp(-tL/to2)-exp(-(tL-dt)/to2)) ) * C * e0 + (exp(-dt/to1)-1) *
alpham1L(:,iel) + (exp(-dt/to2)-1) * alpham2L(:,iel);

        f0(sctrB) = f0(sctrB)-B'*sigma0L(:,iel)*W(kk)*det(J0);

    end                       % end of looping on GPs

    alpham01(:,iel) = alpham1L(:,iel) ;

    alpham02(:,iel) = alpham2L(:,iel) ;

    end                       % end of looping on elements

% *********************************
%    ESSENTIAL BOUNDARY CONDITION
% *********************************

% *********************************
%       SOLUTION OF EQUATIONS
% *********************************

u  = K\ftotal ;
u_x = u(1:2:2*numnode) ;
u_y = u(2:2:2*numnode) ;
utotal = utotal + u ;
utotal_x = utotal(1:2:2*numnode) ;
utotal_y = utotal(2:2:2*numnode) ;

for e = 1 : numelem
    sctr = element(e,:);
    nn   = length(sctr);
    U    = element_disp(e,pos,enrich_node,utotal);
    if (ismember(e,tip_elem)) % tip element
        for inod=1:1
            if (EnodTpSpRef(sctr(inod))==1)
               xTip=xTip1;
               xCr=xCr1;
               alpha=alpha1;
            else
               xTip=xTip2;
               xCr=xCr2;
               alpha=alpha2;
            end
        end
    elseif  ( any(intersect(tip_nodes,sctr)) ~= 0)% having tip enriched nodes
```

(Continued)

BOX 8.1 (Continued)

```
            for inod=1:nn
                if (EnodTpSpRef(sctr(inod))==2)
                    xTip=xTip2;
                    xCr=xCr2;
                    alpha=alpha2;
                    break
                elseif (EnodTpSpRef(sctr(inod))==1)
                    xTip=xTip1;
                    xCr=xCr1;
                    alpha=alpha1;
                    break
                end
            end

        elseif (ismember(e,split_elem))      % split element
                    xTip=xTip2;
                    xCr=xCr2;
                    alpha=alpha2;
        else
            xTip=xTip2;
            xCr=xCr2;
            alpha=alpha2;
        end
        for q=1:nn
            pt = stressPoints(q,:);
            [B,J0] = private_xfemBmatrix4(pt,elemType,e,enrich_node,...
                                   EnodTpSpRef,xCr,xTip,alpha);
            strain = B * U ;

            stress(e,q,:) = (Ebar * C * strain) + sigma0L(:,e) ;

        end

    end

        totalstress(:,:,1)  = totalstress(:,:,1) + stress(:,:,1) ;
        totalstress2(:,:,2) = totalstress2(:,:,2) + stress(:,:,2) ;
        totalstress3(:,:,3) = totalstress3(:,:,3) + stress(:,:,3) ;

    end
```

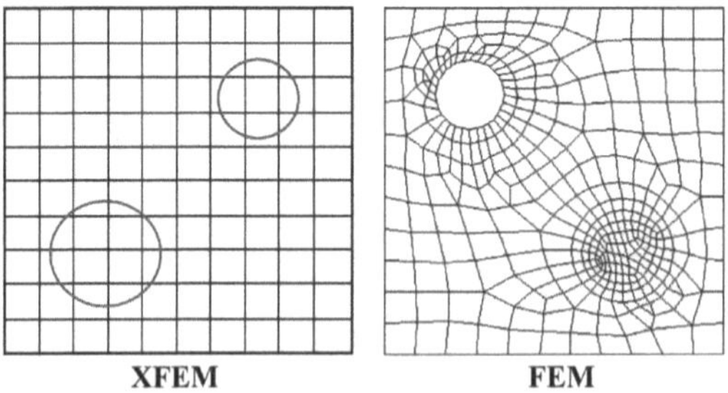

Figure 8.3 Comparison between extended finite element method (XFEM) and finite element method (FEM) mesh generation considering an inclusion within a viscoelastic domain. *From Zhang, H. H., & Li, L. X. (2009). Modeling inclusion problems in viscoelastic materials with the extended finite element method. Finite Elements in Analysis and Design, 45(10), 721–729.*

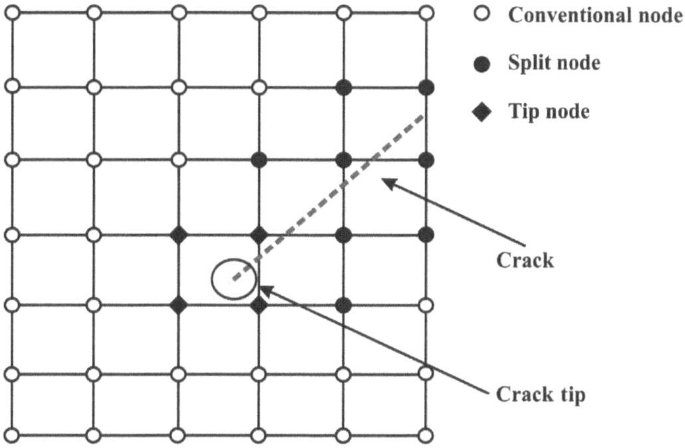

Figure 8.4 Conventional, split, and tip nodes in a meshed domain containing a crack as a discontinuity.

and Φ_K^m ($m = 1, 2, 3$, and 4) represents the local property at the crack tip based on the displacement field around the crack tip as follows (see Section 4.3.4):

$$\left[\Phi_K^1, \Phi_K^2, \Phi_K^3, \Phi_K^4\right] = \left[\sqrt{r}\sin\frac{\theta}{2}, \sqrt{r}\cos\frac{\theta}{2}, \sqrt{r}\sin\theta\sin\frac{\theta}{2}, \sqrt{r}\sin\theta\cos\frac{\theta}{2}\right] \quad (8.73)$$

In Box 8.1 "numnode" represents the total number of nodes which are directly depended on the mesh size and domain area. Two matrices of "split_nodes" and "tip_nodes" represent the nodes of split and tip elements, respectively. According to Eq. (8.71) and considering that this problem is a two-dimensional one, "total_unkown" which must be calculated can be simply specified. Here, we need to define the preallocated stiffness matrix, "*K*," force vector, "*f*," and residual force vector, "*f*0," considering "total_unkown."

To compute stiffness matrix, in the first stage, it is required to specify that the problem must be solved by considering "plane stress" or "plane strain" condition to select the elastic matrix based on Eqs. (8.3) or (8.4). After that, it is required to determine Gauss points and the nodal force vector, which is eliminated here, as both of them are usual sections of a conventional FEM code. Some examples will be presented in this chapter to show how nodal force vector should be defined in some cases.

Then, as can be seen in Box 8.1, a loop is designed for each time step to calculate the equations derived in Section 8.2. The "sigma0" is a vector which contains residual stresses and "alpham" is a vector which is needed for calculating residual stress [see Eqs. (8.22) and (8.23)]. Here, in the presented code, a Prony series with five parameters (one isolated spring with two Maxwell models) is used as an example. It is evident that it will be possible to extend this equation for different configurations of the generalized Maxwell model or the generalized Kelvin–Voigt model. As can be seen in Box 8.1, two loops are used for performing the calculation, the external loop is defined on time steps and the internal loop is defined on elements. The functions "discontQ4quad" and "disTipQ4quad" are used to determine quadrature points and their weights highlighted in the code as [W, Q] which are required for integration. These functions, shown in Boxes 8.2 and 8.3, are used for

BOX 8.2 A function for determining quadrature points and their weights in MATLAB for split elements to use within an extended finite element method framework. ***From Hajikarimi, P. (2013). Numerical Modeling of Healing Characteristics of Asphalt Mixtures, M.Sc thesis, University of Tehran, Tehran, Iran.***

```
function [W,Q] = discontQ4quad(order,phi)

corner = [1 2 3 4 1];
node   = [-1 -1; 1 -1; 1 1; -1 1];

% loop on element edges
for i = 1 : 4
    n1 = corner(i);
    n2 = corner(i+1);
    if ( phi(n1)*phi(n2) < 0 )
        r    = phi(n1)/(phi(n1)-phi(n2));
        pnt  = (1-r)*node(n1,:)+r*node(n2,:);
        node = [node;pnt];
    end
end

tri = delaunay(node(:,1),node(:,2));

tri = tricheck(node,tri);
% loop over subtriangles to get quadrature points and weights
pt = 1;
for e = 1:size(tri,1)
    [w,q]=quadrature(order,'TRIANGULAR',2);
    % transform quadrature points into the parent element
    coord = node(tri(e,:),:);
    a = det([coord,[1;1;1]])/2;
    if ( a<0 )  % need to swap connectivity
        coord = [coord(2,:);coord(1,:);coord(3,:)];
        a = det([coord,[1;1;1]])/2;
    end

    if ( a~=0 )
        for n=1:length(w)
            N=lagrange_basis('T3',q(n,:));
            Q(pt,:) = N'*coord;
            W(pt,1) = 2*w(n)*a;
            pt = pt+1;
        end
    end

end
```

BOX 8.3 A function for determining quadrature points and their weights in MATLAB for tip elements to use within an extended finite element method framework. ***From Hajikarimi, P. (2013). Numerical Modeling of Healing Characteristics of Asphalt Mixtures, M.Sc thesis, University of Tehran, Tehran, Iran.***

```
function [W,Q]=disTipQ4quad(order,phi,nodes,tip)

epsilon = 0.00001;
corner  = [1 2 3 4 1];
node    = [-1 -1; 1 -1; 1 1; -1 1];

coord = zeros(1,2);
ksi   = 0;
eta   = 0;
iter  = 10;

inc = 1;
while (inc < iter)
    [N,dNdxi]=lagrange_basis('Q4',coord);    % compute shape functions

    x = N'*nodes(:,1);
    y = N'*nodes(:,2);
    df1dr = dNdxi(:,1)' * nodes(:,1);
    df1ds = dNdxi(:,2)' * nodes(:,1);
    df2dr = dNdxi(:,1)' * nodes(:,2);
    df2ds = dNdxi(:,2)' * nodes(:,2);

    f1 = x - tip(1);
    f2 = y - tip(2);

    detF = df1dr*df2ds - df1ds*df2dr ;

    invf(1,1) =  1.0/detF * df2ds;
    invf(1,2) = -1.0/detF * df1ds;
    invf(2,1) = -1.0/detF * df2dr;
    invf(2,2) =  1.0/detF * df1dr;

    ksi = ksi - invf(1,1)*f1 - invf(1,2)*f2;
    eta = eta - invf(2,1)*f1 - invf(2,2)*f2;

    coord(1) = ksi;
    coord(2) = eta;

    if( (abs(ksi - coord(1)) < epsilon) && ...
            (abs(eta - coord(2)) < epsilon) )
        inc  = iter + 1;
        ntip = coord;
    else
        inc = inc + 1;
    end
end

% loop on element edges
for i = 1:4
    n1 = corner(i);
    n2 = corner(i+1);
    if phi(n1)*phi(n2) < 0
        r    = phi(n1)/(phi(n1)-phi(n2));
        pnt  = (1-r)*node(n1,:)+r*node(n2,:);
        node = [node;pnt];
    end
end

% insert the tip into the Delaunay triangulation
node = [node;ntip];
% get decomposed triangles
tri = delaunay(node(:,1),node(:,2));
tri = tricheck(node,tri);
```

(Continued)

BOX 8.3 (Continued)

```
% loop over subtriangles to get quadrature points and weights
pt = 1;
for e=1:size(tri,1)
    [w,q] = quadrature(order,'TRIANGULAR',2);
    % transform quadrature points into the parent element
    coord = node(tri(e,:),:);
    a = det([coord,[1;1;1]])/2;
    if ( a<0 )  % need to swap connectivity
        coord=[coord(2,:);coord(1,:);coord(3,:)];
        a = det([coord,[1;1;1]])/2;
    end

    if ( a~=0 )
        for n=1:length(w)
            N = lagrange_basis('T3',q(n,:));
            Q(pt,:) = N'*coord;
            W(pt,1) = 2*w(n)*a;
            pt = pt+1;
        end
    end

end
```

elements containing some enriched nodes or discontinuity. The Delaunay method is used to discretize the considered element to some triangle subsections and then to determine Gauss points for each triangle.

For conventional elements, the function "quadrature" is used, which was developed in the Department of Mechanical Engineering at Northwestern University as presented in Box 8.4. This function quadrature returns a n^*1 column vector W of quadrature weights and a n^*dim matrix of quadrature points, where n is the number of quadrature points. The function is called [W, Q] = quadrature (nint, type, dim). In this function, nint is the quadrature order, type is the type of quadrature (i.e., Gaussian, triangular, etc.) and dim is the number of special dimensions of the problem. The default for type is "GAUSS" and the default for dim is unity.

The strain–displacement matrix, B, introduced in Eq. (8.65) can be represented in the XFEM framework as follows:

$$B = [B_{\text{FEM}} | B_{\text{XFEM}}] \tag{8.74}$$

The nodal parameters are stored in the following manner:

$$u = [u_1 v_1 u_2 v_2 \ldots u_n v_n | a_1 b_1 \ldots a_n b_n] \tag{8.75}$$

BOX 8.4 A function "quadrature" for determining quadrature points and their weights in MATLAB for conventional elements.

```
function [W,Q] = quadrature( quadorder, qt, sdim )

  if ( nargin < 3 )   % set default arguments
    if ( strcmp(qt,'GAUSS') == 1 )
      dim = 1;
    else
      dim = 2;
    end
  end

  if ( nargin < 2 )
    type = 'GAUSS';
  end

  if ( strcmp(qt,'GAUSS') == 1 )

    if ( quadorder > 8 )  % check for valid quadrature order
      disp('Order of quadrature too high for Gaussian Quadrature');
      quadorder =8;
    end

    quadpoint=zeros(quadorder^sdim ,sdim);
    quadweight=zeros(quadorder^sdim,1);

    r1pt=zeros(quadorder,1); r1wt=zeros(quadorder,1);

    switch ( quadorder )
      case 1
        r1pt(1) = 0.000000000000000;
        r1wt(1) = 2.000000000000000;

      case 2
        r1pt(1) = 0.577350269189626;
        r1pt(2) =-0.577350269189626;

        r1wt(1) = 1.000000000000000;
        r1wt(2) = 1.000000000000000;

      case 3
        r1pt(1) = 0.774596669241483;
        r1pt(2) =-0.774596669241483;
        r1pt(3) = 0.000000000000000;

        r1wt(1) = 0.555555555555556;
        r1wt(2) = 0.555555555555556;
        r1wt(3) = 0.888888888888889;

      case 4
        r1pt(1) = 0.861134311594053;
        r1pt(2) =-0.861134311594053;
        r1pt(3) = 0.339981043584856;
        r1pt(4) =-0.339981043584856;

        r1wt(1) = 0.347854845137454;
        r1wt(2) = 0.347854845137454;
        r1wt(3) = 0.652145154862546;
        r1wt(4) = 0.652145154862546;

      case 5
        r1pt(1) = 0.906179845938664;
        r1pt(2) =-0.906179845938664;
```

(Continued)

BOX 8.4 (Continued)

```
    rlpt(3) = 0.538469310105683;
    rlpt(4) =-0.538469310105683;
    rlpt(5) = 0.000000000000000;

    rlwt(1) = 0.236926885056189;
    rlwt(2) = 0.236926885056189;
    rlwt(3) = 0.478628670499366;
    rlwt(4) = 0.478628670499366;
    rlwt(5) = 0.568888888888889;

  case 6
    rlpt(1) = 0.932469514203152;
    rlpt(2) =-0.932469514203152;
    rlpt(3) = 0.661209386466265;
    rlpt(4) =-0.661209386466265;
    rlpt(5) = 0.238619186003152;
    rlpt(6) =-0.238619186003152;

    rlwt(1) = 0.171324492379170;
    rlwt(2) = 0.171324492379170;
    rlwt(3) = 0.360761573048139;
    rlwt(4) = 0.360761573048139;
    rlwt(5) = 0.467913934572691;
    rlwt(6) = 0.467913934572691;

  case 7
    rlpt(1) =  0.949107912342759;
    rlpt(2) = -0.949107912342759;
    rlpt(3) =  0.741531185599394;
    rlpt(4) = -0.741531185599394;
    rlpt(5) =  0.405845151377397;
    rlpt(6) = -0.405845151377397;
    rlpt(7) =  0.000000000000000;

    rlwt(1) = 0.129484966168870;
    rlwt(2) = 0.129484966168870;
    rlwt(3) = 0.279705391489277;
    rlwt(4) = 0.279705391489277;
    rlwt(5) = 0.381830050505119;
    rlwt(6) = 0.381830050505119;
    rlwt(7) = 0.417959183673469;

  case 8
    rlpt(1) =  0.960289856497536;
    rlpt(2) = -0.960289856497536;
    rlpt(3) =  0.796666477413627;
    rlpt(4) = -0.796666477413627;
    rlpt(5) =  0.525532409916329;
    rlpt(6) = -0.525532409916329;
    rlpt(7) =  0.183434642495650;
    rlpt(8) = -0.183434642495650;

    rlwt(1) = 0.101228536290376;
    rlwt(2) = 0.101228536290376;
    rlwt(3) = 0.222381034453374;
    rlwt(4) = 0.222381034453374;
    rlwt(5) = 0.313706645877887;
    rlwt(6) = 0.313706645877887;
    rlwt(7) = 0.362683783378362;
    rlwt(8) = 0.362683783378362;

  otherwise
    disp('Order of quadrature to high for Gaussian Quadrature');
```

(Continued)

BOX 8.4 (Continued)

```
    end  % end of quadorder switch

    n=1;

    if ( sdim == 1 )
      for i = 1:quadorder
        quadpoint(n,:) = [ r1pt(i) ];
        quadweight(n) = r1wt(i);
        n = n+1;
      end

    elseif ( sdim == 2 )
      for i = 1:quadorder
        for j = 1:quadorder
          quadpoint(n,:) = [ r1pt(i), r1pt(j)];
          quadweight(n) = r1wt(i)*r1wt(j);
          n = n+1;
        end
      end

    else % sdim == 3
      for i = 1:quadorder
        for j = 1:quadorder
          for k = 1:quadorder
            quadpoint(n,:) = [ r1pt(i), r1pt(j), r1pt(k) ];
            quadweight(n) = r1wt(i)*r1wt(j)*r1wt(k);
            n = n+1;
          end
       end
     end

    end

    Q=quadpoint;
    W=quadweight;
  % END OF GAUSSIAN QUADRATURE DEFINITION

  elseif ( strcmp(qt,'TRIANGULAR') == 1 )

    if ( sdim == 3 )  %%% TETRAHEDRA

      if ( quadorder ~= 1 &&  quadorder ~= 2 &&  quadorder ~= 3  )
        % check for valid quadrature order
        disp('Incorect quadrature order for triangular quadrature');
        quadorder = 1;
      end

      if  ( quadorder == 1 )
        quadpoint = [ 0.25 0.25 0.25 ];
        quadweight = 1;

      elseif ( quadorder == 2 )
        quadpoint = [ 0.58541020  0.13819660  0.13819660;
                      0.13819660  0.58541020  0.13819660;
                      0.13819660  0.13819660  0.58541020;
                      0.13819660  0.13819660  0.13819660];
        quadweight = [1; 1; 1; 1]/4;

      elseif ( quadorder == 3 )
        quadpoint = [ 0.25  0.25  0.25;
                      1/2   1/6   1/6;
                      1/6   1/2   1/6;
                      1/6   1/6   1/2;
                      1/6   1/6   1/6];
```

(Continued)

BOX 8.4 (Continued)

```
        quadweight = [-4/5 9/20 9/20 9/20 9/20]';

      end

      Q=quadpoint;
      W=quadweight/6;

    else  %%% TRIANGLES

      if ( quadorder > 7 ) % check for valid quadrature order
        disp('Quadrature order too high for triangular quadrature');
        quadorder = 1;
      end

      if ( quadorder == 1 )   % set quad points and quadweights
        quadpoint = [ 0.3333333333333, 0.3333333333333 ];
        quadweight = 1;

      elseif ( quadorder == 2 )
        quadpoint = zeros( 3, 2 );
        quadweight = zeros( 3, 1 );

        quadpoint(1,:) = [ 0.1666666666667, 0.1666666666667 ];
        quadpoint(2,:) = [ 0.6666666666667, 0.1666666666667 ];
        quadpoint(3,:) = [ 0.1666666666667, 0.6666666666667 ];

        quadweight(1) = 0.3333333333333;
        quadweight(2) = 0.3333333333333;
        quadweight(3) = 0.3333333333333;

      elseif ( quadorder <= 5 )
        quadpoint = zeros( 7, 2 );
        quadweight = zeros( 7, 1 );

        quadpoint(1,:) = [ 0.1012865073235, 0.1012865073235 ];
        quadpoint(2,:) = [ 0.7974269853531, 0.1012865073235 ];
        quadpoint(3,:) = [ 0.1012865073235, 0.7974269853531 ];
        quadpoint(4,:) = [ 0.4701420641051, 0.0597158717898 ];
        quadpoint(5,:) = [ 0.4701420641051, 0.4701420641051 ];
        quadpoint(6,:) = [ 0.0597158717898, 0.4701420641051 ];
        quadpoint(7,:) = [ 0.3333333333333, 0.3333333333333 ];

        quadweight(1) = 0.1259391805448;
        quadweight(2) = 0.1259391805448;
        quadweight(3) = 0.1259391805448;
        quadweight(4) = 0.1323941527885;
        quadweight(5) = 0.1323941527885;
        quadweight(6) = 0.1323941527885;
        quadweight(7) = 0.2250000000000;

      else
        quadpoint = zeros( 13, 2 );
        quadweight = zeros( 13, 1 );

        quadpoint(1 ,:) = [ 0.0651301029022, 0.0651301029022 ];
        quadpoint(2 ,:) = [ 0.8697397941956, 0.0651301029022 ];
        quadpoint(3 ,:) = [ 0.0651301029022, 0.8697397941956 ];
        quadpoint(4 ,:) = [ 0.3128654960049, 0.0486903154253 ];
        quadpoint(5 ,:) = [ 0.6384441885698, 0.3128654960049 ];
        quadpoint(6 ,:) = [ 0.0486903154253, 0.6384441885698 ];
        quadpoint(7 ,:) = [ 0.6384441885698, 0.0486903154253 ];
        quadpoint(8 ,:) = [ 0.3128654960049, 0.6384441885698 ];
        quadpoint(9 ,:) = [ 0.0486903154253, 0.3128654960049 ];
        quadpoint(10,:) = [ 0.2603459660790, 0.2603459660790 ];
```

(Continued)

BOX 8.4 (Continued)

```
            quadpoint(11,:) = [ 0.4793080678419, 0.2603459660790 ];
            quadpoint(12,:) = [ 0.2603459660790, 0.4793080678419 ];
            quadpoint(13,:) = [ 0.3333333333333, 0.3333333333333 ];

            quadweight(1 ) = 0.0533472356088;
            quadweight(2 ) = 0.0533472356088;
            quadweight(3 ) = 0.0533472356088;
            quadweight(4 ) = 0.0771137608903;
            quadweight(5 ) = 0.0771137608903;
            quadweight(6 ) = 0.0771137608903;
            quadweight(7 ) = 0.0771137608903;
            quadweight(8 ) = 0.0771137608903;
            quadweight(9 ) = 0.0771137608903;
            quadweight(10) = 0.1756152576332;
            quadweight(11) = 0.1756152576332;
            quadweight(12) = 0.1756152576332;
            quadweight(13) =-0.1495700444677;

          end

          Q=quadpoint;
          W=quadweight/2;
        end

      end  % end of TRIANGULAR initialization

    % END OF FUNCTION
```

As can be seen in Eq. (8.75), the additional unknowns are placed at the end of the vector. Hence, the true nodal displacement is simply defined as $u\_x = u$(1:2:2*numnode) and $u\_y = u$(2:2:2*numnode).

Here, to determine the strain–displacement matrix, a function "private_xfemBmatrix4" is used presented in Box 8.5.

The code is continued by implementing recursive equations to determine the stiffness matrix. Essential boundary conditions must be introduced to the code for each problem individually. After that, the solution of the equation can be simply performed with the stiffness matrix and nodal force vector. In the end, the stresses are computed by using the derived equations for residual stress [see Eq. (8.23)]. In the following section, some examples are presented and solved using the incremental form of viscoelastic relations and developed numerical framework for a viscoelastic medium.

BOX 8.5 A function for determining strain–displacement matrix in MATLAB to use within an extended finite element method framework. ***From Hajikarimi, P. (2013). Numerical Modeling of Healing Characteristics of Asphalt Mixtures, M.Sc thesis, University of Tehran, Tehran, Iran.***

```
function [B,J0] =
private_xfemBmatrix4(pt,elemType,e,enrich_node,EnodTpSpRef
                               ,xCr,xTip,alpha)
global node element

sctr = element(e,:);
nn   = length(sctr);
[N,dNdxi] = lagrange_basis(elemType,pt(:,1:2)); % element shape functions
J0 = node(sctr,:)'*dNdxi;                       % element Jacobian matrix

invJ0 = inv(J0);
dNdx  = dNdxi*invJ0;                            % derivatives of N w.r.t XY
Gpt = N' * node(sctr,:);                        % GP in global coord, used
QT  =[cos(alpha) sin(alpha); -sin(alpha) cos(alpha)];

% Bfem is always computed
Bfem = zeros(3,2*nn);
Bfem(1,1:2:2*nn)   = dNdx(:,1)' ;
Bfem(2,2:2:2*nn)   = dNdx(:,2)' ;
Bfem(3,1:2:2*nn)   = dNdx(:,2)' ;
Bfem(3,2:2:2*nn)   = dNdx(:,1)' ;

% Switch between non-enriched and enriched elements
if ( any(enrich_node(sctr)) == 0 ) % Non-enriched elements
    B = Bfem;
else                               % Enriched elements
    Bxfem = [] ;
    % loop on nodes, check node is enriched ...
    for in = 1 : nn
        if ( enrich_node(sctr(in)) == 1)     % H(x) enriched node
            % Enrichment function, H(x) at global Gauss point
            dist = signed_distance(xCr,Gpt);
            if (cos(alpha) <0)
                dist=-dist;
            end
            Hgp  = sign(dist);
            % Enrichment function, H(x) at node "in"
            dist = signed_distance(xCr,node(sctr(in),:));
            if (cos(alpha) <0)
                dist=-dist;
            end
            Hi    =   sign(dist);
            % Bxfem at node "in"
            BI_enr = [dNdx(in,1)*(Hgp - Hi) 0 ;
                0 dNdx(in,2)*(Hgp - Hi) ;
                dNdx(in,2)*(Hgp - Hi) dNdx(in,1)*(Hgp - Hi)];
            % Add to the total Bxfem
            Bxfem = [Bxfem BI_enr];
            clear BI_enr ;
        elseif ( enrich_node(sctr(in)) == 2) % B(x) enriched node
            % compute branch functions at Gauss point
            xp     = QT*(Gpt-xTip)';            % local coordinates

            r      = sqrt(xp(1)*xp(1)+xp(2)*xp(2));
            theta = atan2(xp(2),xp(1));
            if ( theta > pi || theta < -pi)
                disp (['something wrong with angle ',num2str(thet)]);
            end
            [Br,dBdx,dBdy] = private_branch(r,theta,alpha);

            % compute branch functions at node "in"
            xp     = QT*(node(sctr(in),:)-xTip)';
            r      = sqrt(xp(1)*xp(1)+xp(2)*xp(2));
```

(Continued)

BOX 8.5 (Continued)

```
                theta = atan2(xp(2),xp(1));
                if ( theta > pi || theta < -pi)
                    disp (['something wrong with angle ',num2str(thet)]);
                end
                [BrI] = branch_node(r,theta);

                % composants of B_enr matrix

                aa = dNdx(in,1)*(Br(1)-BrI(1)) + N(in)*dBdx(1) ;
                bb = dNdx(in,2)*(Br(1)-BrI(1)) + N(in)*dBdy(1) ;
                B1_enr = [aa 0 ; 0 bb ; bb aa];

                aa = dNdx(in,1)*(Br(2)-BrI(2)) + N(in)*dBdx(2) ;
                bb = dNdx(in,2)*(Br(2)-BrI(2)) + N(in)*dBdy(2) ;
                B2_enr = [aa 0 ; 0 bb ; bb aa];

                aa = dNdx(in,1)*(Br(3)-BrI(3)) + N(in)*dBdx(3) ;
                bb = dNdx(in,2)*(Br(3)-BrI(3)) + N(in)*dBdy(3) ;
                B3_enr = [aa 0 ; 0 bb ; bb aa];

                aa = dNdx(in,1)*(Br(4)-BrI(4)) + N(in)*dBdx(4) ;
                bb = dNdx(in,2)*(Br(4)-BrI(4)) + N(in)*dBdy(4) ;
                B4_enr = [aa 0 ; 0 bb ; bb aa];

                BI_enr = [B1_enr B2_enr B3_enr B4_enr];
                clear B1_enr; clear B2_enr; clear B3_enr; clear B4_enr;
                Bxfem = [Bxfem BI_enr];
                clear BI_enr ;
            end
        end         % end of loop on nodes
        % B matrix
        B = [ Bfem Bxfem ];
        clear Bfem; clear Bxfem;
    end             % end of switch between enriched and non-enriched elements
```

8.5 Numerical examples

In this section, two different examples are presented to better understand how incremental form of viscoelastic relations can be used to solve problems by implementing XFEM. The first example is an infinite viscoelastic plate with a circular hole under uniaxial tension and the second example is an edge crack in an infinite viscoelastic plate.

8.5.1 Infinite viscoelastic plate with a circular hole under uniaxial tension

A viscoelastic quadrangle plate with length and width of $T = 2$ and a circular void in the center with a radius of $a = 0.4$ is considered, as illustrated in Fig. 8.5. The considered medium is a part of an infinite viscoelastic plate where a creep load of σ_∞ is applied on its boundaries in the x-direction. Applying such a creep load results in a displacement on the plate's boundaries which can

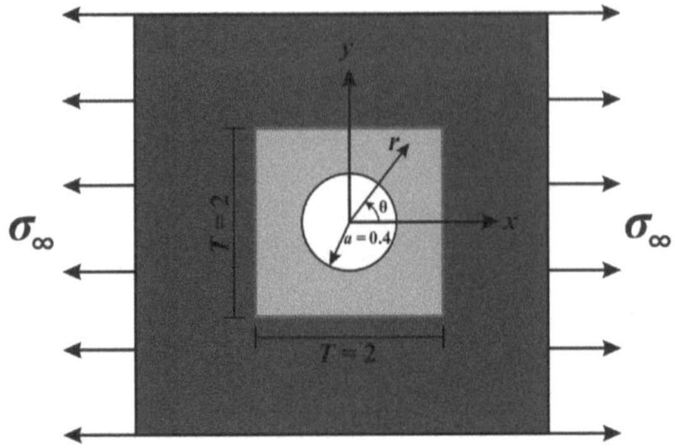

Figure 8.5 An infinite viscoelastic plate with a circular void subjected to uniaxial tension.

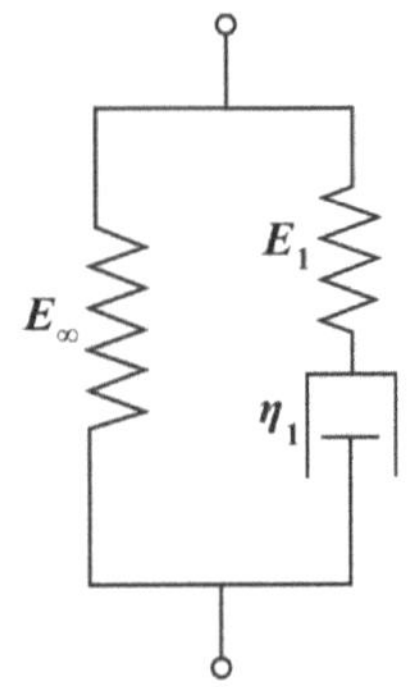

Figure 8.6 The simple three-parameter standard viscoelastic model.

be determined from the analytical solution for the displacement field of an infinite viscoelastic plate by implementing the correspondence principle. A simple three-parameter standard viscoelastic model is used to explain the viscoelastic behavior of this plate (see Fig. 8.6) in which $E_\infty = 1.0$, $E_1 = 1.0$, and $\eta_1 = 10.0$. The Poisson's ratio is considered to be a constant value of 0.3.

By implementing the correspondence principle introduced in Chapter 4, Correspondence Principle of Viscoelasticity, the exact displacement field can be derived for an infinite viscoelastic plate as follows:

$$u_x = \frac{\sigma_\infty(1+\nu)\vartheta(t)a}{4}\left\{\frac{r}{a}(\kappa+1)\cos\theta + 2\frac{a}{r}[(\kappa+1)\cos\theta + \cos3\theta] - 2\frac{a^3}{r^3}\cos3\theta\right\} \tag{8.76}$$

$$u_y = \frac{\sigma_\infty(1+\nu)\vartheta(t)a}{4}\left\{\frac{r}{a}(\kappa-3)\sin\theta + 2\frac{a}{r}[(1-\kappa)\sin\theta + \sin3\theta] - 2\frac{a^3}{r^3}\sin3\theta\right\} \tag{8.77}$$

in which a is the radius of circular void and (r,θ) is the polar coordinate in which the origin coincides with the center of the hole. Since plane strain condition is considered in this problem, $\kappa = 3 - 4v$. Then, considering a three-parameter standard viscoelastic model, the creep compliance would be:

$$D(t) = \frac{1}{E_\infty} - \frac{E_1}{E_\infty(E_\infty + E_1)} \exp\left(-t/\tau'\right) \tag{8.78}$$

where

$$\tau' = \left(\frac{1}{E_\infty} + \frac{1}{E_1}\right)\eta_1 \tag{8.79}$$

The exact displacements calculated by using analytical solution presented in Eqs. (8.76) and (8.77) are applied on boundaries of considered finite plate having creep load of $\sigma_\infty = 0.25$.

By considering four different mesh sizes, the problem is solved for 200 s loading time and time step of $\Delta t = 5$ s to investigate the effect of element size on the results. The meshes are illustrated in Fig. 8.7 which

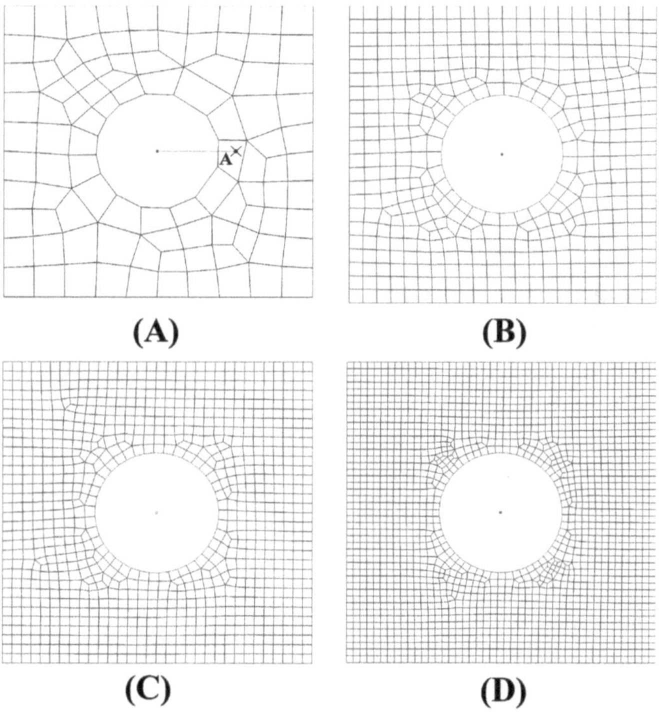

Figure 8.7 Four different mesh sizes of a viscoelastic plate with a circular void including (A) 97, (B) 438, (C) 907, and (D) 1796 elements.

have 97, 438, 907, and 1796 elements generated by implementing AutoMesh2D-one Domain software.

The displacement of Point A ($r = 0.5$, $\theta = 0.0$) which is shown in Fig. 8.7 is determined to show the accuracy of problem solving. This displacement is calculated based on both analytical solution and FEM's solution to compare the results. Considering the material properties and main assumptions of this problem, the analytical solution for both u_x and u_y would be:

$$u_x = 0.139035(1 - 0.5\exp(-t/10)) \tag{8.80}$$

$$u_y = 0 \tag{8.81}$$

To determine the displacement at Point A in x and y direction, it is required to use FE approximation. Fig. 8.8 shows a four-node element in $x-y$ coordinates and its mapped form in the $\xi-\eta$ coordinate system. Based on FEM, it is possible to determine $x^T = (x, y)$ by interpolating nodal values of $\bar{\boldsymbol{x}}^T = (\bar{x}, \bar{y})$ as follows:

$$x = \sum_{j=1}^{4} N_j \bar{\boldsymbol{x}}_j \tag{8.82}$$

in which N_j is shape function matrix, which is expressed as:

$$N_j = \begin{bmatrix} N_j & 0 \\ 0 & N_j \end{bmatrix} \tag{8.83}$$

In a finite isoperimetric element, the displacement field of $u^T = (u_x, u_y)$ can be determined by interpolating nodal displacements by using the following equation:

$$u = \sum_{j=1}^{4} N_j \bar{\boldsymbol{u}}_j \tag{8.83}$$

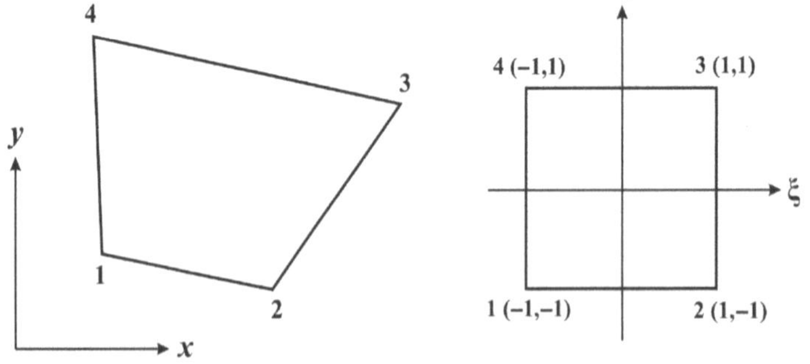

Figure 8.8 The isoparametric finite element.

To determine u at a point in the $x-y$ coordinate system based on nodal displacement of the element containing this node, it is required to have the coordinate of the correspondent point in the $\xi-\eta$ coordinate system. For the elements in a structured mesh, it is so simple to find the corresponding point while for the case represented in this example, an iterative procedure must be used. Fig. 8.9 shows displacement of Point A during a loading time of 200 s for different mesh sizes. As can be seen in this figure, the solution of this problem is sensitive to the number of elements and converges to an analytical solution for 1000 elements. The vertical displacements at Point A are zero due to pure tension loading and symmetry condition.

It is also required to investigate the effect of time step size on the results. For this purpose, four different time steps are considered, including $\Delta t = 20$ s, $\Delta t = 10$ s, $\Delta t = 5$ s, and $\Delta t = 1$ s. Then, the problem is solved by using 2000 elements. Fig. 8.10 depicts results of numerical modeling in comparison with an analytical solution. As can be seen in this figure, changing the time step size has no significant effect on the results and it can be concluded that the linear viscoelastic solution is a stable solution.

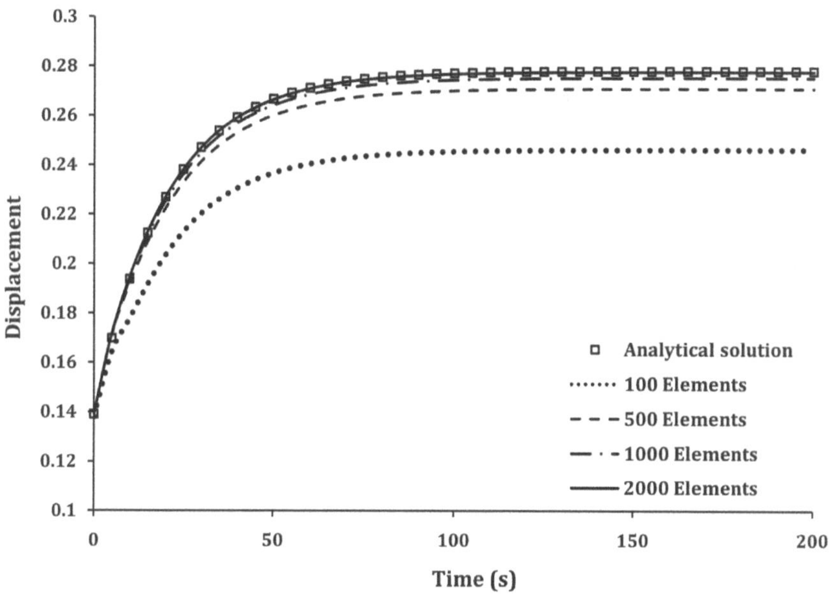

Figure 8.9 Results of displacement versus time for 200 s loading time with four different mesh sizes for a viscoelastic plate with circular void.

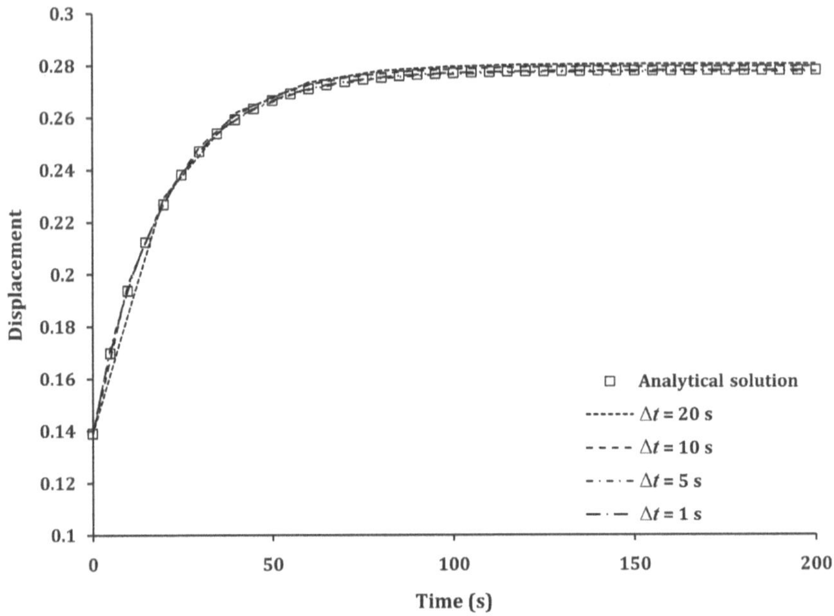

Figure 8.10 Results of displacement versus time for 200 s loading time with four different time steps for a viscoelastic plate with circular void.

8.5.2 The first and second opening mode for an edge crack with in-plane behavior

In this example, a finite viscoelastic plate is considered which is separated from an infinite viscoelastic plate and loaded by an analytical displacement field on its boundaries, as depicted in Fig. 8.11. As can be seen in this figure, the plate has an edge crack. Therefore it is required to use the XFEM code presented in Section 8.4.

The viscoelastic behavior of the considered plate is expressed by using three parameters shown in Fig. 8.6 in which $E_\infty = 1.0$, $E_1 = 1.0$, and $\eta_1 = 10.0$. The Poisson's ratio is considered to be a constant value of 0.3. In Fig. 8.10, $L = 2$ and crack length $a = 1$ are selected. The loading time is 200 s and the time step $\Delta t = 0.5$ s. The displacement field around the finite plate is determined for mode I and mode II of the fracture by considering $K_I(t) = 0.5$ and $K_{II}(t) = 0.25$. In Section 4.3.4, the analytical solution for two-dimensional linear viscoelastic fracture mechanics is derived for determining the displacement field around the crack tip for both mode I (plane crack opening) and mode II (in-plane shear) of the

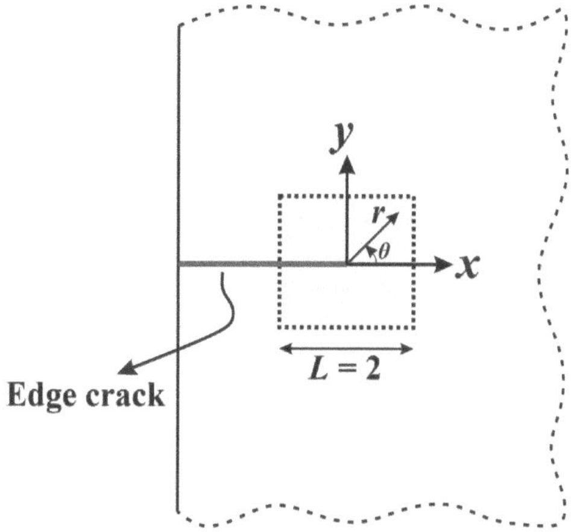

Figure 8.11 An edge crack within an infinite viscoelastic plate for simulating mode I and mode II of fracture.

fracture. Here, the displacement field is again presented [see Eqs. (4.72) and Eqs. (4.73)] as:

$$u_x(r,\theta,t) = (1+\nu)\phi(t)K_I(t)\sqrt{\frac{r}{2\pi}}\cos\frac{\theta}{2}(\kappa - \cos\theta) \tag{8.84}$$

$$u_y(r,\theta,t) = (1+\nu)\phi(t)K_I(t)\sqrt{\frac{r}{2\pi}}\sin\frac{\theta}{2}(\kappa - \cos\theta) \tag{8.85}$$

For the mode I, and:

$$u_x(r,\theta,t) = (1+\nu)\phi(t)K_{II}(t)\sqrt{\frac{r}{2\pi}}\sin\frac{\theta}{2}(\kappa + \cos\theta + 2) \tag{8.86}$$

$$u_y(r,\theta,t) = -(1+\nu)\phi(t)K_{II}(t)\sqrt{\frac{r}{2\pi}}\cos\frac{\theta}{2}(\kappa + \cos\theta - 2) \tag{8.87}$$

For the second mode of fracture, in which:

$$\phi(t) = \frac{1}{E_\infty} - \frac{E_1}{E_\infty(E_\infty + E_1)}\exp(-t/\tau') \tag{8.88}$$

and,

$$\tau' = \left(\frac{1}{E_\infty} + \frac{1}{E_1}\right)\eta_1 \tag{8.89}$$

and $\kappa = 3 - 4\upsilon$ for plane strain condition of this problem.

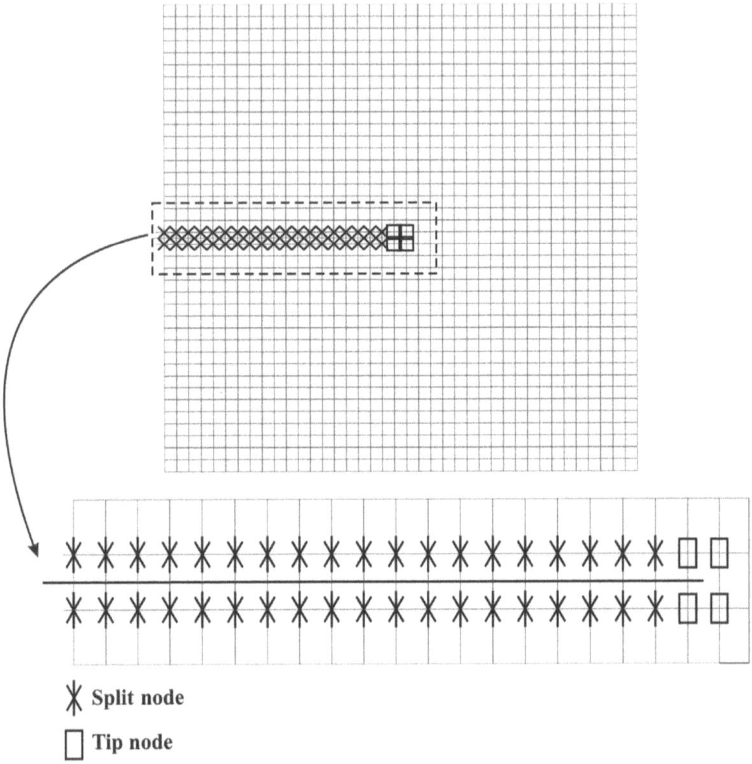

Figure 8.12 The structured mesh used in the extended finite element method framework for modeling an edge crack in an infinite viscoelastic plate.

To solve this problem, a structured 39×39 mesh (see Fig. 8.12) is considered and the displacement fields calculated based on analytical solution of mode I and mode II are applied on its boundaries. Box 8.6 shows the coded added to the main MATLAB code for applying analytical solution on boundaries of the finite part as external loading. The deformed shape of mesh after applying the load for $t = 200$ s is shown in Fig. 8.13 for both mode I and mode II of fracture.

As can be seen in Fig. 8.12, the elements are categorized as: (1) conventional elements, (2) split elements, and (3) tip elements. To enrich the nodes and involve the analytical solution near the crack tip into problem solution, some additional degrees of freedom are considered for split and tip elements, as shown in Fig. 8.12. The Heaviside function is

BOX 8.6 The commands added to the main MATLAB code for applying analytical solution on boundaries as external loading.

```
for ii=1:nny

    LNodes (ii) = ( ii -1 ) * nny + 1 ;

    RNodes (ii) = ii * nny ;

end

BoundCond = unique ( [botNodes;RNodes';topNodes;LNodes'] ) ;

BoundNodes = node (BoundCond,:) ;

for ee = 1:size(BoundCond,1)

    bound = BoundCond(ee);
    boundx = (bound * 2) - 1 ;
    boundy = bound * 2 ;

    radius = sqrt(abs( ((BoundNodes(ee,1))^2) ) + abs( ((BoundNodes(ee,2))^2) )) ;

    th = atan2 (BoundNodes(ee,2),BoundNodes(ee,1)) ;

    ud(boundx) = ud(boundx) + (1+nu) * (Jdisp1-Jdisp0) * sqrt (radius/2/pi)
                 * (kapa - 1 + 2 * ((sin(th/2))^2)) * cos(th/2) ;

    ud(boundy) = ud(boundy) + (1+nu) * (Jdisp1-Jdisp0) * sqrt (radius/2/pi)
                 * (kapa + 1 - 2 * ((cos(th/2))^2)) * sin(th/2) ;
end

f = -(K * ud)  ;


ftotal = f + f0 ;
```

implemented for modeling the discontinuity within the split elements (with × sign) and the enrichment function represented in Eq. (8.73) is used for the tip element (with □ sign). The degrees of freedom for enriched nodes of split elements which is shown with × is 2 and the degrees of freedom for enriched nodes of the tip element which is shown with □ is 8.

To verify the XFEM solution, two points of A ($r = 0.5$, $\theta = \pi$) and B ($r = 0.5$, $\theta = -\pi$) are considered, which are in the same position in the Cartesian coordinate system on the middle of crack but are above and below the crack in the polar coordinate system. Using such a property, it is possible to determine crack opening displacement (COD) in mode I

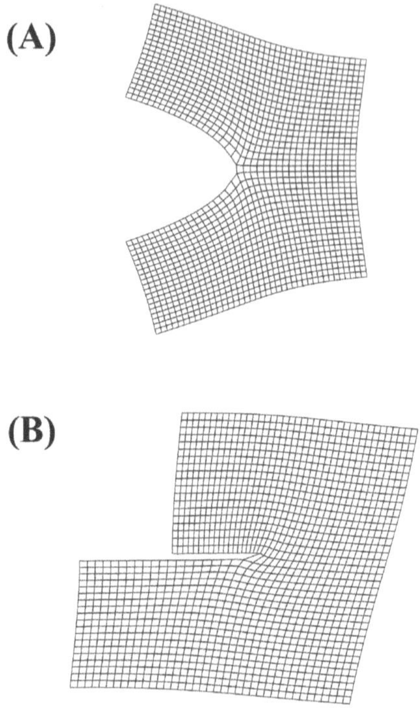

Figure 8.13 Deformed shape of finite part of an infinite viscoelastic plate in presence of an edge crack for (A) mode I and (B) mode II of fracture.

and crack sliding displacement (CSD) in mode II based on the following equations:

$$\mathrm{COD} = u_{y,I}(r, \pi, t) - u_{y,I}(r, -\pi, t) \tag{8.90}$$

$$\mathrm{CSD} = u_{x,II}(r, \pi, t) - u_{x,II}(r, -\pi, t) \tag{8.91}$$

The XFEM approximation for COD and CSD is as follows:

$$\mathbf{u}^h(\mathbf{x}) = \sum_{j=1}^{n} N_j(\mathbf{x})\,\mathbf{u}_j + \sum_{k=1}^{m} N_k(\mathbf{x})\big(H(\xi) - H(\xi_k)\big)\,\mathbf{a}_k \tag{8.92}$$

in which $H(x) = 1$ if x is above the crack and $H(x) = -1$ if x is below the crack. The j index is for conventional degrees of freedom and k is for enriched ones. Figs. 8.14 and 8.15 show COD and CSD of both analytical and numerical solutions which show the ability of XFEM for solving

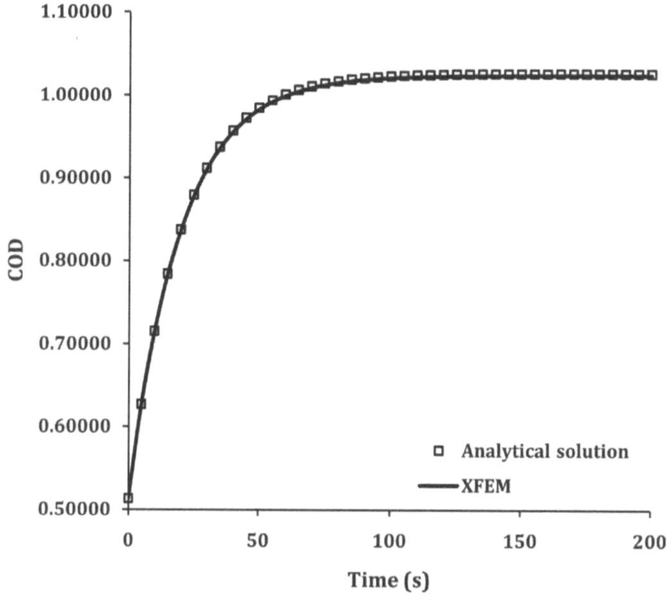

Figure 8.14 Analytical and numerical results of crack opening displacement (COD) for mode I of fracture for an edge crack within an infinite viscoelastic plate. *XFEM*, Extended finite element method.

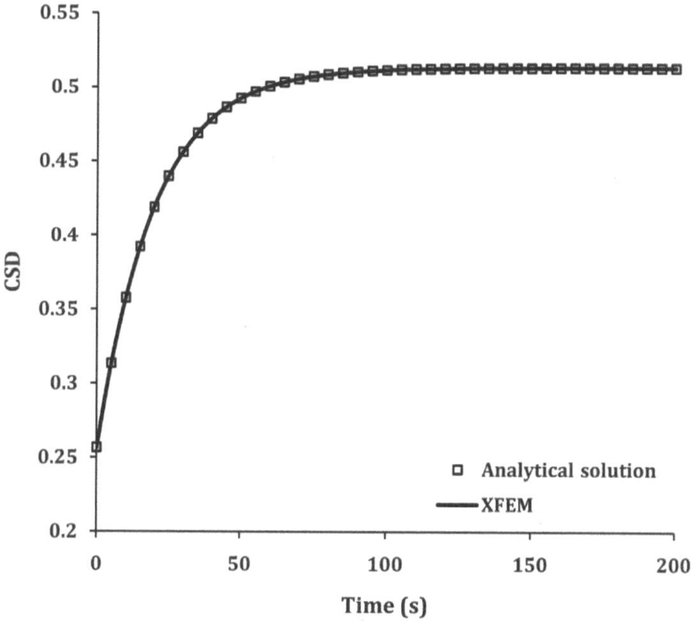

Figure 8.15 Analytical and numerical results of crack sliding displacement (CSD) for mode II of fracture for an edge crack within an infinite viscoelastic plate. *XFEM*, Extended finite element method.

viscoelastic problems containing discontinuity. Hajikarimi et al. (2012) verified a similar problem by considering a five-parameter generalized Maxwell model.

References

Banerjee, P. K., & Butterfield, R. (1981). *Boundary element methods in engineering science* (Vol. 17). London: McGraw-Hill.

Barber, J. R. (2002). *Elasticity*. Springer.

Chessa, J., Smolinski, P., & Belytschko, T. (2002). The extended finite element method (XFEM) for solidification problems. *International Journal for Numerical Methods in Engineering*, *53*(8), 1959–1977.

Hajikarimi, P., Aflaki, S., & Hoseini, A. S. (2013). Implementing fractional viscoelastic model to evaluate low temperature characteristics of crumb rubber and gilsonite modified asphalt binders. *Construction and Building Materials*, *49*, 682–687. Available from: https://doi.org/10.1016/j.conbuildmat.2013.09.001.

Hajikarimi, P., Mohammadi, S., & Aflaki, S. (2012). Two dimensional creep analysis of a linear cracked viscoelastic medium using the extended finite element method. In: *ECCOMAS 2012–European Congress on Computational Methods in Applied Sciences and Engineering*, E-Book Full Papers.

Hajikarimi, P., Onochie, A., & Fini, E. H. (2020). Characterizing mechanical response of bio-modified bitumen at sub zero temperatures. *Construction and Building Materials*, *240*, Article 117940. Available from: https://doi.org/10.1016/j.conbuildmat.2019.117940.

Liu, G.-R., & Gu, Y.-T. (2005). *An introduction to meshfree methods and their programming*. Springer Science & Business Media.

Moës, N., Dolbow, J., & Belytschko, T. (1999). A finite element method for crack growth without remeshing. *International Journal for Numerical Methods in Engineering*, *46*(1), 131–150. Available from: https://doi.org/10.1002/(SICI)1097-0207(19990910)46:1 < 131::AID-NME726 > 3.0.CO;2-J.

Mohammadi, S. (2008). *Extended finite element method: For fracture analysis of structures*. John Wiley & Sons.

Mohammadi, S. (2012). *XFEM fracture analysis of composites*. John Wiley & Sons Incorporated.

Reddy, J. N. (1993). *An introduction to the finite element method* (p. 27) New York: McGraw-Hill.

Sadd, M. H. (2009). *Elasticity: Theory, applications, and numerics*. Academic Press.

Soutas-Little, R. W. (2012). *Elasticity*. Courier Corporation.

Zhang, H. H., & Li, L. X. (2009). Modeling inclusion problems in viscoelastic materials with the extended finite element method. *Finite Elements in Analysis and Design*, *45*(10), 721–729. Available from: https://doi.org/10.1016/j.finel.2009.06.006.

Zienkiewicz, O. C., & Morice, P. B. (1971). *The finite element method in engineering science* (Vol. 1977). London: McGraw-Hill.

Index

Note: Page numbers followed by "*f*," "*t*," and "*b*" refer to figures, tables, and boxes, respectively.

R

S

T

FSC
www.fsc.org
MIX
Papier | Fördert gute Waldnutzung
FSC® C083411

Kolibri 360 GmbH
Ferdinand-Jühlke-Straße 7
99095 Erfurt, Deutschland
produktsicherheit@kolibri360.de